WALDEN III: A CATHOLIC AMERICA

WITH A PERSPECTIVE FROM KURT GODEL

by

Patrick O'Dougherty

Published by the Hellenist America Institute: Minneapolis, 1991

The Hellenist America Institute Publishing Company
Riverside Plaza M3410
1615 South Fourth Street
Minneapolis, Minnesota 55454

Library of Congress Catalog Card Number:

Printed in the United States of America

Library of Congress Cataloging-in-Publication Data

O'Dougherty, Patrick Aquinas, 1946--
     A new Catholic statement in America: history and
     mathematics
     Includes index

ISBN
     1. A Catholic America. 2. Kurt Godel--Mathematics and History. 3. Catholic
Radicalism. 4. Apologists. 5. Catholic Intellectual History. 6. The Power Structure of the
Church in America. 7. Integralism.

Dedication

To Richard Kast and Patricia O'Dougherty-Kast. To Megan and Sean and Stephan. To Mary
Boggie and David Noble. To a Catholic America. To Mike.

Table of Contents

"History is not was, but is."
Barbara Fields quoting William Faulkner

Introduction

The purpose of this book is to present a new case for a Catholic America. There are many different ways that American history could have developed. For example, America was initially a theocracy. It could have continued as a theocracy. Also, America could have become a confederacy, or a proletariat nation, or a German or French speaking nation. Can America become radically different? Yes, it could become a Catholic nation. What could a Catholic nation do for the American way of life? If the Church had a greater say the bureaucracy could be liquidated and more social and economic justice could be delivered to the poor. Racism and nationalism could also be greatly decreased because the Church is opposed to these ideas. The thesis of this book is that "mathematics is the underlying reality of the universe."[1] Mathematics is the key to a Catholic America. All of American history is open to abstraction and quantification. Mathematics is the hint towards the nature of God's mind. Behind the objective "stuff"[2] of the universe and American history is mathematics. To make this topic more manageable, the writer is going to focus on the religious and philosophical implications of Kurt Godel's mathematic's proof. His is the idea that there are some things in mathematics and also in life that are true but can't be proven to be true. Method is math. The writers chosen to present and develop the case for or picture of a Catholic America are first, Daniel and Philip Berrigan, radical priests; second, Henry Thoreau, the philospher; third, the Catholic Catechism; fourth, C.S. Lewis, the Apologist; fifth, Father E. J. Cuskelly, the spiritual director; sixth, Margaret Reher, the historian; seventh, Father Powell's book on abortion; and finally, T.J. Reese's book, Archbishop. Mathematic proofs, like Godel's proof, give a philosophical legitimacy to Catholicism and American history. The writers called to present and defend the legitimacy of a Catholic America are the matrix. New approaches to Catholicism such as fissionism, fusionism and field theory are going to be presented. Is an "underlying reality" of these approaches mathematics? Mike Franey might argue this. Is Godel's proof a paradox? The writer thinks that this is true. Is a key to the Bible mathematics? Yes. Will Christianity unite? It is possible. Solidarity, concord, cooperation, fellowship, harmony, unity, firmness, reliability, soundness, and stability instead of competition are the themes to build a Catholic America. Catholicism is integral to America. Americans should unite themselves in a philosophy of Catholic integralism, that is, completeness. Christ is the whole. America should be a Catholic integral.

A CATHOLIC AMERICA--A MATHEMATICAL PERSPECTIVE

Godel

It was an accepted conclusion that each area of mathematical thought can be equipped with a number of axioms that would enable a mathematician to generate systematically a large number of true propositions about a chosen area of mathematical research. Godel's proof indicates that this theory is untenable. The implication of his work is that the axiomatic method is limited. For example, arithmetic cannot ever be completely axiomatized. Godel proved that it is not possible to develop the logical "consistency" of a large area of a deduction system like arithmetic. Thus no complete systemization of many critical areas of mathematics is possible.[3]

The "impossibility of deducing the parallel axiom from others was demonstrated." "A proof can be given of the impossibility of proving certain propositions within a system." Godel's thesis is a proof of the impossibility of demonstrating "certain important propositions in arithmetic."[4]

Is mathematics still "the science of quantity?" Today mathematics is the field that derives its conclusions from a given set of axioms or theorems. The pure mathematician derives the "necessary logical consequences" of initial assumptions. For example, the idea of class does not prove the "consistency of any particular system built on it." "Cantor's theory of infinite classes" is pertinent. He argues that "N is normal if and only if N is non normal." Thus the theory that N is normal and non normal is both true and false.[5]

Models are used to solve problems. David Hilbert tried to concoct "absolute" proofs, wherein the consistency of a mathematical system could be developed without taking for granted the consistency of some other mathematical system. Hilbert contended that the first move to develop an absolute proof "is the complete formalization of a deductive system." This means first pruning the expressions that occur within the system "of all meaning." The signs become empty signs. Precise rules determine how the signs are to be manipulated. "A system of signs" is developed which hides nothing and contains only those things which we put in it.[6]

The theorems and postulates of this formal system are "strings" or marks without meaning; and, they are constructed by rules for combining the signs of the "calculus" into larger systems. Moreover, when a system has been entirely formalized, the derivation of theorems from postulates is like the transformation of one set of "strings" into another set of "strings."[7]

A sheet composed of "meaningless" signs like those mentioned above doesn't assert much. It is an abstract design whose structure is determined. They are like statements filled with meaning "about a meaningful system: and do not belong to that particular system. They are a language "about" mathematics. Hilbert called them "meta-mathematics." "The description, discussion, and theorizing about the systems belong in the file marked "meta-mathematics."[8]

Hilbert's idea of proof was to show by finite means the impossibility of deriving various contradictory formulas in a set of mathematical calculus or "meta-chess." Bertrand Russell's Principia, sets for an instrument for studying the whole system of arithmetic as a system of marks without meaning "an uninterpreted calculus."[9]

Godel "showed that it is impossible to give a meta-mathematical proof of the consistency of a system comprehensive enough to contain the whole of arithmetic." "Godel showed that the Principia, or any other system within which arithmetic can be developed is essentially incomplete. Thus, "given any consistent set of arithmetical axioms, there are true arithmetical statements that cannot be derived from the set."[10]

Hilbert mapped geometry onto algebra. Using the idea of a mirror or mirroring, Godel arithmetizes meta-mathematics. That is, "each meta-mathematical statement is represented by a unique formula within arithmetic; and, the relations of logical dependence between meta-mathematical concepts are fully reflected in the numerical relations of dependence between their corresponding arithmetical formulas." Godel "showed that G is demonstrable if, and only if, its formal negation is demonstrable.[11]

"If a formula and its own negation are both formally demonstrable, the arithmetical calculus is not consistent." "If the calculus is consistent, G is a formally undecidable formula." "Therefore, if arithmetic is consistent, G is formally undecidable." "Godel then proved that though G is not formally demonstrable, it is a true arithmetical formula." "Since G is both true and formally undecidable, the axioms of arithmetic are incomplete."[12]

"Godel's analysis should not be misunderstood; it does not exclude a meta-mathematical proof of the consistency of arithmetic." "What it excludes is a proof of consistency that can be mirrored by the formal deduction of arithmetic.[13]

Godel and a Mathematics System

What is a mathematical system? It is a "set of assumptions and the set of conclusions that can be proved from them by DEDUCTION." To prevent circular reasoning an axiom is chosen. An axiom is a "statement that is assumed to be true without proof." The conclusions of a mathematical system are labelled theorems. There are many different mathematical systems "including

arithmetics: real number arithmetic; clock or modular, arithmetic; transfinite arithmetic; arithmetic without numbers
geometries: Euclidean geometry; spherical geometry; other non-Euclidean geometries;
algebra: algebra with real numbers; matrix algebra; vector algebra; algebra of sets."

4

A CATHOLIC AMERICA--A MATHEMATICAL PERSPECTIVE

Each of these mathematical systems has its own axioms.[14]

What is an approach delineating the "foundations of mathematical systems?" How many axioms are enough? Usually, an explanation of axioms creates more questions? For example, attempts to show that Euclid's parallel postulate was "an unnecessary axiom for geometry led to the discovery of non-Euclidean geometries with different fifth postulates."[15]

Two aims of mathematical systems are <u>completedness</u> and <u>consistency</u>. That is, a set of axioms must be complete and not have contradictions. For example, consider any statement (A) and its negation (not A):

"In a <u>complete</u> system it is possible to prove either A or its negation.

In a <u>consistent</u> system it is not possible to prove both A and its negation."[16]

As was mentioned earlier, in the first part of the twentieth century, David Hilbert, a German mathematician, attempted to define a list of axioms for the arithmetic of whole numbers; and, then he tried to show that "two contradictory statements could not both be derived from his set of axioms." If he had been able to do this the axioms would be proven consistent. He failed in his attempt.[17]

In 1931, Kurt Godel, pointed out that consistency for a true axiomatic system like Hilbert's cannot be proved relying only on the resources of that system. Hilbert desired his system to be complete as well as consistent. Therein lies the problem. A paradox is produced. A paradox is "a case of apparently valid deductive reasoning that leads to a contradiction of a fact or premise." For example, a paradox "arises when you consider the sentence which was considered before in the language of arithmetic that says essentially

"This statement is not provable."

"If this statement is false, then you can prove it using axioms, but that means the system is inconsistent. If the statement is true, then you cannot prove it and the system is incomplete."[18]

"Godel showed that Hilbert's system cannot be both consistent and complete." "Such a system is then called <u>incomplete</u>, and the question of consistency is called <u>undecidable</u>."[19]

What are the implications of Godel's proof for America or for Catholic American history? Maybe, like arithmetic, America or American history is incomplete. Maybe America is undecidable. Also, maybe the idea of God is true but not always provable. Moreover, if there is a system or language that can be developed about mathematics called meta-mathematics, maybe there is a system of mathematics that can be developed about behavior called meta-behavior or meta-biology.

Can some truths about Catholicism be reduced to arithmetic or algebra or

A CATHOLIC AMERICA--A MATHEMATICAL PERSPECTIVE

geometry. Plato might think so, if we might include mathematic ideas along with his ideas or ideals. Is the "underlying reality" of America "mathematics"? Franey might think so. Can behavior be abstractly described by mathematics? The answer is yes. Can genetics be analyzed by a mathematical model? The answer is yes. Is history mathematics? Yes. Is Catholic history meta-history? The answer is yes. This is one of several theses for the rest of this book.

A CATHOLIC AMERICA--A MATHEMATICAL PERSPECTIVE

American Catholic Radicalism

Thesis: Radicalism is a Class or Set of Behaviors. Radicals are a group.

May 17, 1968

In the Tradition of Thoreau

During May 17, 1968 a political event arose that was critical as a large affront to the legitimacy of the present day nation state which supported it. On the day of May 17, 1968 two highly visible American clergymen and their cohorts burned a large amount of the Selective Service files of Catonville, Maryland, while saying "killing is disorder, life and gentleness, and community and unselfishness are the only order we recognize. For the sake of that order we risk our liberty our good name."[1]

The clergymen responsible for this political act of civil disobedience were the Berrigan brothers, Daniel and Philip. Because of this act and the prison terms which resulted and their writings, these priests brought about the largest challenge in the modern world to the idea that one can be a faithful Christian and an obedient citizen at the same moment in time.[2]

What was the Berrigan's brothers view of the nature of the "contemporary" nation and specifically, the American nation? What was their conception of the duty of the Christian in such a nation? How did they view human history? Why did these clergymen leave their churches for the napalm firing of U.S. government Civil Service records?[3]

The Berrigan's on Nation State Ethics

Philip used remarkable Biblical passages to describe his attitudes. For example, while thinking about Satan's third temptation of Jesus while he was on his retreat in the desert, where Satan promised Christ "all the kingdoms of the world in their majesty" if he would worship Satan. Christ answered Satan by saying "Worship the Lord your God and serve only Him."' Philip Berrigan points out that the Devil would not display what was not his.[4]

Daniel Berrigan does not mean a full condemnation of all nations. "Only a minority of Believers he says "are living under states which could be labelled as sinful." That is, they are in principle in opposition to belief in God and religious practices.[5]

These priests are agreed on the nature of the modern nation state and current times political theory. Philip contends that "Western society...stands today as the enemy of man." Today we are in times during which only a new beginning would alleviate the evil of the nation state, America. America, according to Daniel Berrigan is "corrupt and evil."[6]

What are the sins of America? There are four groups. First, American foreign policy is not charitable. Americans use their power to hold back the power of the underdeveloped nations. For example, the Vietnam war was genocidal.[7]

Second, America's policy of total war as the main policy of our foreign affairs is gravely evil. The black people and the poor constitute almost twenty-five per cent of the American population. They are "technocracies' victims."[8]

Third, America maintains negroes in a colonial status. According to the Berrigans, it is not just several of the nation's policies that are flawed, it is the central ethic of America--our greed and materialism. Philip says that we must choose " between God and wealth and America." For example, our nation's drive for more and more consumer products brings about a type of internal war. A battle field is the inevitable "other end of the consumer process." One cannot "be a servant of God and of society at the same time." International power plays are immoral. Daniel Berrigan argues that the Bible makes illegal the use of evil to fight evil. Daniel Berrigan contends that religion should dominate over politics. He opposes the Church's just war theory. He says that in "the U.S.A. we want it different or not at all."[9]

Fourth, the problem of church and state can be sinful because the state is a "foreign power" to Christians. My dad, James O'Dougherty, claimed that the state is the "original, original sin." My thesis on the Church and state is that the state is a non sequitur. It does not follow legitimately from the Church.[10]

Do Not's for the Christian

Do not have anything to do with a state's war efforts. Do not pay taxes to a government's war effort. Christian resistance is not "just disobedience because a Christian should oppose leading a 'normal' existence in a society such as the United States." The only choice for the Christian is to be a " revolutionary;" "Go underground or die!" The selective service should be put out of business through nonviolence. We need a spiritual revolution like Christ's revolution. The Berrigans try to justify some violence for oppressed groups like the Black Panthers and Angela Davis.[11]

The Berrigans call upon us to take on a "negative" stance. We are called to take a civil disobedient approach that would break up immoral laws, while avoiding violence to individuals. Daniel Berrigan is opposed to liberalism because the warfare state is enslaving us. He contends that the Church has been bought off by state governments. The Church should take on a protest role.[12]

In his book on Dante, The Discipline of the Mountain, Daniel Berrigan argues that politics either left or right offers little alternative. Both sides are a waste. He raises the question, Is Dante right? That is, is life a laborious rise to God and to sanctity? Are we living in a" world on the make"? Does Dante come into his own in his Purgatorio? Daniel Berrigan places himself, Dante and mankind in Purgatory.

A CATHOLIC AMERICA--A MATHEMATICAL PERSPECTIVE

Daniel decides to ascend into Dante's mountain of the dead to find knowledge. Dante is his leader.[13]

The pilgrims leave Easter morning to climb the "Mount of Purgation." As the odyssey is undertaken, they gain the "Portal of Purgation." On the "Terrace of Pride," the strengths of humility are opened to them. On the "Terrace of Envy" they are enveloped in a fog; and, on the "Terrace of Sloth" the two pilgrims are chided. An earthquake befalls them. On the "Terrace of Avarice" the plight of the Christian is shown to them. On the "Terrace of Gluttony" the discipline of their adventure is shown to them. On the "Terrace of Lust" mankind's lust is purged by fire. Finally, in the earthly paradise the somewhat cruel climb of the Mount of Purgation is "consummated."[14]

What is Daniel Berrigan getting at here? What he is trying to do is to set Dante's Purgatorio in the nuclear warfare world. His activist stance is a discipline like Dante's climb of the mountain of Purgatory.

The Future: Destruction or Perfection?

According to the Berrigans, we are near the point of a quantum leap in evolution. We should try through resistance to purify the world. They argue that God's greatest achievement is the perfection of resistance communities of man. The Berrigans reject original sin. Man, they contend, can be perfected. The end of original sin is genetics and more sin. Daniel and Philip accepted reservations about their rejection of original sin after they were released from prison.[15]

Augustine: "The State as Irrelevant to Salvation"

In contrast to the Berrigans, Augustine had a belief in original sin as the focus of his theology. He maintained the position that the bulk of mankind won't avoid the effects of this sin through redemption by God's saving grace. We can't expect much reform. The state isn't the tool of reform. The state isn't relevant to salvation. The Roman State was the world's greatest robber. The citizen must obey his particular state. Soldiers can be forced to fight in unjust wars. Charity is the reason that Augustine urges us to defend the state. He rejects the Berrigans' activism. The Christian should remain detached in his relationship with the state. Man has a fallen nature. The "City of Man" is pride. The "City of God" is adoration. Augustine had a doctrine of predestination wherein God predestined few for salvation and the remaining members will be damned.[16]

The State in Augustine's Philosophy

Reform won't come through the state. The classical view of man is that the state can help secure perfection. One loves with emotions "not with the mind." The

9

state does not have a role in man's perfection or reformation. The state exists to perpetuate peace and contain disorder. The state often fails in justice. Complete justice can only be found in the "City of God." States are like bank robbers. The city of the ungodly lacks complete justice. Augustine equates the Catholic Church with the "City of God." He contends that the residents of the "City of God" will only be a small portion of all people alive at a particular time. Thus it would be unlikely that a Christian state would arise. Nations are relatively unjust.[17]

Morales and Kingdoms and Augustine

The Roman Empire was composed mainly by members of the "City of Man." This state was held together by a need for the acquiescence of human desires and a scarcity of goods. Augustine criticizes the Roman State and its empire. He contends that an empire fails to bring about joy. The material fails to bring joy to people. He allowed defensive battles. The Berrigans, in contrast, condemn violence even when it is utilized in a nation's protection.[18]

War and Peace

Augustine argues that wars and empires are necessary because both men and nations are evil and sinful. Sin brings about government; and, war is a necessary wrong. Man has a political obligation to cooperate with the attempts of the state to achieve a common good even if this means that some people must die.[19]

Augustine and the Christian and his Obligation to the State

Augustine maintains that the government has some right to our bodies. It doesn't have a right to own our souls. Disobedience is only rarely tolerated. The Christian should obey the nation. Augustine argues that a Christian can remain guiltless in obeying an "unrighteous order to fight." He maintains that Christians have an obligation to take upon themselves public duty and office.[20]

In contrast, the Berrigans believe that the perfectibility of man's community is God's noblest work. Augustine maintained the opposite view. He was negative about man's perfectibility. The Berrigans feel that the poor can be sanctified as a group. In contrast, Augustine maintains that the best way for a Christian to love his neighbor is by having him love God. He wants the Christian to focus on the "City of God."[21]

Reinhold Niebuhr

Is Niebuhr for championing power politics, a nation's interest, and the present? Or, is he a liberal Christian who favors creating the Spirit of the Divine on earth? He is not a political quietest like St. Augustine or a political activist like the Berrigans. Niebuhr is dedicated to both the human and the spiritual. He was a socialist until 1953. Niebuhr's thought went through two stages. During the first stage he was an adherent to the pacifism of the Social Gospel. In the second stage he was a Marxist. During this period he rejected great distinctions to be made between violence and nonviolence. Niebuhr rejects the modern optimistic orientation towards man.[22]

What is Niebuhr's conception of original sin? It is the sin of pride, he thinks. One example of this is man's attempts to deny his mortality. Man often seeks to become completely divine. Man's move to abuse his freedom and to push to the limit his own power is his original sin. Man is a creature rather than a creator. He is a creature that sins. His guilt implies free will. His recognition of his sins is his greatest idea of freedom.[23]

Niebuhr on Man's Future

Christianity's claims can't be verified. Faith justifies man. The world is a "vale of tears." "Man's hope" is forgiveness, Christ's resurrection of his body, and heaven. Niebuhr does not believe in Christ's Divinity. He rejects a "City of God" after death. He thinks that Christ can't be finite and infinite at the same time. Man can't create a utopia.[24]

Man's Morality

Man is judged by agape or sacrificial love. It is the norm. Niebuhr emphasizes human freedom. In contrast, the Berrigans argue for redemption in this world. Augustine emphasizes the next world. Niebuhr does not like the idea of natural law because he doesn't trust reason. He does not try to specify the nature of moral law that can be understood apart from God's revelation. Mutual love is man's greatest achievement thinks Niebuhr.[25]

American Democratic Morality

Niebuhr does not expect a high level of morality from nations or groups. Group morality often sins against pride. The relationships between states must usually be political rather than ethical. A person can be a good citizen and a good Christian simultaneously.[26]

Nations and their Moral Character

Nations have "prudent self-interest." But, Niebuhr thinks, it is wrong for a country to act on a rule of complete self-interest. Nations should be ruled by agape. In short, a nation should be ruled by a heightened sense of morality and self-interest.[27]

Niebuhr and Vietnam

Viet Nam was, according to Niebuhr, a completely just war. He was opposed to unilateral withdrawal. Niebuhr was opposed to using the bomb on Japan.[28]

World Government

Niebuhr rejects the idea of world government. It would be unstable. It may, however, evolve. A world government would have a world army. Presently, he notes, we have the nation-state grouping.[29]

War

Niebuhr spurns pacifism. It is not responsible behavior. Pacifists are "parasites." Wars should be transformed into crusades.[30]

Niebuhr's Social Teachings: Love, Equality, and Justice

Justice is about relationships. It is the placing of agape into society. "Love is the motive; but, justice is the instrument." Human justice is not a constant; it is relative. He favors equality as an ideal. Man is unusually free.[31]

Radicalism

Niebuhr is a liberal. He advocates incremental change as opposed to radical change or a great reaction. He favors resistance to authority. The powerless, he thinks, should have their power increased. However, he isn't opposed to worldwide revolutionary movements.[32]

Niebuhr on Augustine and Political Realism

"Realism" indicates a position of taking all variables in a social, economic and political arena into account. "Idealism" is noticeable by adherence to moral rules and ideals as opposed to self-interest. Augustine was one of the first "realists" in Western Civilization. He thinks that the "City of Man" is the product of a "tenuous truce between warring factions." Thus Augustine is a realist in regards to government.[33]

Augustine describes the idea of a primitive society rule thus: "This is the prescribed order of nature." Man has dominion over the plants and animals of the earth. Man is rational and he does not dominate over other men; but, he rules over beasts. Love not egocentrism is the rule of man's existence. Niebuhr points out that

A CATHOLIC AMERICA--A MATHEMATICAL PERSPECTIVE

Augustine thinks that the "City of Man" and the "City of God" should mingle.[34]

Christianity and Poverty

Niebuhr thinks that the Church should be on the side of the poor. However, he lacks a true faith in Christ. He maintains that Christianity should be focused on both God and man.[35]

A Criticism of the Berrigans

The Berrigans hope for heaven on earth. They advocate redemption politics. The state should serve as a vehicle for man's redemption. In contrast, Augustine brands the state with the sign of evil. He doesn't think that the state can deliver justice. Can man be perfected? The Berrigans say yes. They reject violence. Augustine, in contrast, contends that when God bans killing he also intends to oppose the sins that follow killing like hatred. Thus killing is not completely wrong and neither is war. Niebuhr argues that the best way for international politics to go is to follow a standard of enlightened self-interest. Thus the policies of other states should also be considered. Niebuhr's outlook does not come to terms with the relationship of Christianity and international politics. Nations act differently than individuals do. An example of Niebuhr's "enlightened self-interest" is the Marshall Plan. In contrast, Augustine contends that a nation should pursue peace.[36]

The Berrigans claim that no person can be a soldier. We shouldn't pay taxes for a war. Christians should actively oppose war efforts. In contrast, Niebuhr focuses on the individual's conscience. Augustine sees that the soldier is an instrument. He thinks that a man can be a good soldier, a good citizen and a good Christian simultaneously.[37]

The Berrigans contend that a Christian should be different. Augustine argues that Christianity does not necessarily mean pacifism. Niebuhr contends that the Christian must pursue justice. Augustine doesn't feel that the pursuit of justice is the critical element for the Christian. He doesn't reject it, however. Human nature keeps a society from becoming very just. Niebuhr argues that both justice and liberty are important. He likes the idea of progress. Niebuhr also thinks that there is not paradise in history--there are no "Christian politics."[38]

To sum up, Niebuhr, in contrast to the Berrigans, is for the idea of original sin and for the idea of the self-interest of the nation state. It seems clear to the writer that there is another position to be taken with regard to original sin and activism. It is to be for both ideas.

A serious limitation of the thoughts of the Berrigans, Niebuhr, and Augustine is that they don't look at the lot of women and children in the world. The point of view of feminism is not prevalent in their writings. Shouldn't women try to ward off the effects of original sin and shouldn't they try also to take an activist stance in the world.

How do the Berrigans stand up with other Catholic thinkers and with the Pope? Basically, it seems that the Berrigans have a lot of good ideas about an activist stance in life. However, because they reject original sin their thought is probably classified as a modernist heresy. The modernist thinkers reject the relevancy of original sin and its impact on man's actions. They seek scientific models for man's behavior and thoughts. Often they portray a guiltless field for man.

Another point that it seems relevant to make about the Berrigans is that they are trying to develop a Catholic counter-culture here in America. They want America to be "different or not at all." Their goal seems to be to want to radicalize Christianity.[39]

The Berrigans and Godel

Is pacifism true yet unprovable given the dogmas of Christianity? Perhaps! Are the Berrigans consistent? Are the Berrigans vegetarians? Can Godel be applied to the Christian intellectual synthesis? The writer thinks so. For example, "given any consistent set of axioms" on Christianity "there are true statements that cannot be derived from the set." Christianity may be true but not completely provable. Faith is needed. Meta-history is based on faith.[40] There are undecidable propositons in history and life.

A CATHOLIC AMERICA--A MATHEMATICAL PERSPECTIVE

Walden I

Thoreau likes the motto "that government is best which governs least." He argues "that government is best which governs not at all." Government is an expedient. Thoreau contends that "all governments are sometimes inexpedient." The arguments brought against have a "standing" army can be used against having a "standing" government. The American government lacks the driving force of a single man. Why? Because a single man can shape the direction of a government by changing his will. The government does not keep people free. It does not develop the frontier and it does not educate. Rather American character does that. Is there a government which majorities do not decide right and wrong? Instead, conscience makes this decision. Why do majorities rule the longest? Because they are the strongest.[1]

Everyman has a conscience. Men should not let a legislator or legislators control their consciences. "We should be free men first and a nation's subjects later." We should do what is right first. A corporation lacks a conscience; however, a corporation of conscientious businessmen is a corporation "with a conscience."[2]

The bulk of men serve the state through the military. This is an abomination. Soldiers have the same value as horses and farm animals. The man who gives only part of himself to the people is called a philanthropist. If a man gives his whole self to mankind, he seems to be selfish and of little use.[3]

How did Thoreau's contemporaries act to their present government? Thoreau maintains that the citizens of his day could only be disgracefully associated with his government because of slavery. Governments like machines have friction. For example, when robbery and oppression dominate in a government, then the machine should be stopped. Therefore, when a substantial part of the population, a sixth, are slaves, then men with a conscience should halt the machine of government.[4]

Voting is a form of gamboling or "gamework." The character of the voter is not at stake. Usually voting for a right doesn't do much for the right. A good or wise man will not let the right action to chance nor will a wise man let what is right prevail through majority rule. There is not much virtue in the masses of man's actions. When the majority of men vote to abolish slavery, then there will probably be little slavery to contend with.[5]

The American man has become an "Odd Fellow." That is, he has become gregarious while lacking intellect or self-reliance. While he may help widows and orphans, he might live by the aid of a Mutual Insurance Company which will bury him properly.[6]

To act from principle is revolutionary and it divides families, states, and religions. It also separates the person by dividing the evil in him from the divine. Why don't men cherish the wisdom of the minority? Why does government always crucify the Christ or excommunicate the Luther or the Copernicus and declare Washington and Franklin rebels?[7] The answer, my friend, "is blowing in the wind."[8] It is not readily available.

Is wrong a part of the necessary friction of the governmental machine? Is the Constitution evil? Thoreau thinks, yes, and backs the abolitionists. Thus is there a single man in the state of Massachusetts who ceases to enslave black people and withdraws from the partnership with the state of Massachusetts, and becomes locked up in the prison system, then slavery would eventually abolished.[9]

What should the tax-gatherer or other public official do? He should, according to Thoreau, resign. It is a peaceful revolution not to pay taxes. If the public officials do not resign, then conscience should bleed.[10]

Thoreau and Civil Disobedience

Thoreau paid no poll-tax for a period of six years. He was jailed because of this for a single night. Thoreau didn't feel confined. He felt that the state is not armed with greater worth but with superior force. "If a plant cannot live according to its nature, it dies, and so a man," Thoreau contended. He had a good time in prison. He defended his actions. Thoreau gave a lot of time to the problem of how the American Constitution sanctions slavery. He said that "no man with a genius for legislation has appeared in America." Thoreau pointed out that the New Testament has been written for 1800 years; yet, where is the politician who has the talent to shed what the bible has shed on the science of law making.[11]

A CATHOLIC AMERICA--A MATHEMATICAL PERSPECTIVE

Walden II: B.F. Skinner

B.F. Skinner, the psychologist, wrote a book called, <u>Walden II</u>.[1] It is a novel about an experimental psychological community that conditions a dissenter through principles of behaviorism into becoming a behaviorist. In a tape on B.F. Skinner by Robert Stone, the theories of behaviorism are delineated. In Stone's presentation, Skinner contends that "behavior is lawful" and that we can predict and control behavior. Stone points out that Skinner is not concerned with "invisible" things like "thinking, feeling, dreaming and emotions." Skinner, Stone contends, is concerned with environmental factors, variables, behavior, and "operant conditioning." The theme Skinner follows is, according to Stone, that behavior changes result from various environmental changes. An example that Stone gives is food deprivation. Rats are deprived of nourishment for "10,6, or 2 hours." This deprivation is the independent variable. The dependent variable is the time that it takes for the rats to run through a rat maze. The results are that the longer the rats are deprived of food, the slower the rats run through the maze.[2]

Operant conditioning is different from classical conditioning, that is, the conditioning of reflexes. It deals with large scale behavior like eating or learning a language. These forms of behavior increase after reinforcement Skinner maintains. Negative reinforcement is something different from punishment. It is, Skinner contends, a variable that is taken away. The example that Stone gives is shutting a door to keep out cold air. Writing behavior is increased as a result. Punishment, in contrast, is something added like an electric shock. To extinguish a behavior a reinforcer is removed. The frequency or time schedule of reinforcement, also, affects behavior Skinner finds. Stone uses the example of intermittent reinforcement which can produce a behavior such as gamboling. An example of a fixed ratio schedule of reinforcement is giving a rat a food pellet after pressing a bar four times in a Skinner box. A variable schedule of reinforcement, for example, would be changing or varying the number of times such as, first, after nine minutes and, second, after five minutes that a rat pressing a bar in a Skinner box is fed. Stone lists the primary reinforcers as "food, water, sex, and warmth." He gives the example of money as a secondary reinforcer. Sometimes there is "stimulus generalization." This occurs when a pigeon which has been reinforced for pecking a red colored dot out of three other colored dots. The pigeon begins pecking a similar color like an orange colored dot when given a new set of colored dots to differentiate among. This example is Stone's.[3]

What is behavior shaping? Behavior shaping is, according to Skinner, the reinforcement of successive approximations of behavior. An example that Stone uses to illustrate is reinforcing a child who is learning how to write when the child makes an"o" or a"t". Different behavior segments are reinforced.[4]

In sum, Skinner is opposed, says Stone, to the ideas of the id, ego, unconscious, feelings, or other "invisible" forces. Skinner thinks that we should be

A CATHOLIC AMERICA--A MATHEMATICAL PERSPECTIVE

scientific about our observations about behavior and look at reaction time, shaping, frequency, and operant reinforcement.[5] These observations can be quantified statistically.

A CATHOLIC AMERICA--A MATHEMATICAL PERSPECTIVE

A Catechism is Like a Map or Mirror.

An American Catechism Quoted

THE SACRAMENTS

1. What is a sacrament? A sacrament is an outward sign and effective instrument of God's grace and man's faith.
2. How are the sacraments related to faith? They not only presuppose faith, but, by words and objects, they also nourish, strengthen, and express it.
3. Is the Church a sacrament? By her relationship with Christ, the Church is a kind of sacrament of intimate union with God, and of the unity of all mankind, that is, she is a sign and an instrument of such union and unity.
5. How many specific sacraments are there? There are seven sacraments: (1) baptism, (2) confirmation, (3) penance, (4) holy eucharist, (5) holy orders, (6) matrimony, and (7) anointing of the sick.[1]

BAPTISM

1. What is Baptism? Baptism is the sacrament of rebirth as a child of God, sanctified by the Holy Spirit, uniting the soul with the death and resurrection of Jesus Christ, cleansing from original and personal sins, and welcoming into the community of the Church.
2. What are the effects of the sacrament of baptism? By Baptism the soul is cleansed from original and personal sins, is welcomed into the community of the Church, is permanently related to God, and joined to the priestly, prophetic, and kingly works of Jesus Christ.
3. Is baptism necessary for salvation? Baptism, at least of interpretative desire, is necessary for salvation, for Jesus Christ declared: '"Unless a man be born again of water and the Spirit he cannot enter the kingdom of God."'[2]

CONFIRMATION

1. What is confirmation? Confirmation is the sacrament by which a baptized person receives the seal of the Holy Spirit as preparation for the witness of a mature Christian life.
9. What preparation is necessary for receiving confirmation?
(a) One must be in the state of grace and be fully instructed in the principal doctrines of the Catholic faith.
(b) One must select a Christian name different from one's baptismal name; the bishop uses this new name in confirming.
(c) One must arrange for a sponsor as in baptism, if possible the same one. The

sponsor must be a practicing Catholic who has been confirmed.[3]

PENANCE

1. What is the sacrament of penance? It is the sacrament which brings to Christians God's merciful forgiveness for sins committed after baptism, and reconciles them to the Church, wounded by their sins.
3. Did Jesus Christ give his Church the power to forgive sins? St. Peter and the other apostles were given the power of "binding and loosing" sins.
4. Why is it necessary to confess to have sins forgiven? It is necessary because the forgiving is a judicial pronouncement: it requires evidence which only the penitent can give, if he is able.
5. Do all Catholics go to confession? All Catholics--pope, bishops, priests, and lay people--go to confession.
7. What must we do to obtain forgiveness of our sins in the sacrament of penance? We must do five things: (1) prepare by making an examination of conscience; (2) have sorrow for our sins; (3) resolve never again to commit sins; (4) confess our sins to the priest; (5) perform the penance which the priest assigns.[4]

Indulgences

2. What is an indulgence? An indulgence is the Church's special intercession with God for remission of the temporal punishment due to sins, the guilt of which has already been wiped out.
3. How does the Church have the power to grant indulgences? "The Church, making use of her power of ministering the redemption of Christ our Lord...authoritatively intervenes to dispense to the faithful who are rightly disposed the treasury of satisfaction of Christ and of the saints, for the remission of temporal punishment."[5]

THE HOLY EUCHARIST

1. What is the sacrament of the holy eucharist? The holy eucharist is the chief sacrament; under the appearance of bread and wine there is present the humanity of Jesus Christ, united with his divine person.
5. How did Jesus Christ institute the holy eucharist? "While they were at supper, Jesus took bread, and blessed and broke, and gave it to his disciples, and said, 'Take and eat; this is my body.' And taking a cup, he gave thanks and gave it to them, saying, 'All of you, drink of this; for this is my blood of the new covenant, which is being shed for many unto the forgiveness of sins."' (Matt. 26-28).
8. What is this change of the bread and wine into the body and blood of Jesus called? It is called "transubstantiation" which means "change of substance.

10. Why did Jesus Christ institute the holy eucharist? "He did this in order to perpetuate the sacrifice of the cross throughout the centuries until he should come again."

11. Did Jesus Christ give his apostles the power to change bread and wine into his body and blood? Yes, he gave them that power when he said, "Do this in remembrance of me."

12. How is this power exercised today? This power given by Jesus Christ at the Last Supper is exercised today by the bishops and priests of the Catholic Church in the holy sacrifice of the mass.[6]

THE SACRIFICE OF THE MASS

1. What is the mass? The mass is the sacrifice in which Jesus Christ, through the ministry of priests, perpetuates the sacrifice of the cross by his real presence under the appearances of bread and wine.

2. Is the mass just a memorial ritual? No, the mass is much more than a mere memorial. Jesus Christ is really present in his risen body, continually offering himself to the Father through the ministry of the priest.

6. When was the sacrifice of the mass instituted? At the Last Supper when our Lord told his apostles to do what he had done, namely, to change bread and wine into his body and blood: "Do this in remembrance of me."

9. How does the mass differ from the sacrifice on the cross? The difference is in the manner in which the sacrifice is offered; on the cross our Lord really suffered and died; in the mass it is the risen body of Christ that is present and is offered, which suffers now no more.

11. At what part of the mass do the body and blood of Jesus Christ come upon the altar? At the consecration, when the priest repeats the words of Jesus Christ: "This is my body; this is the cup of my blood."[7]

HOLY COMMUNION

1. What is holy communion? Holy Communion is the receiving of Jesus Christ in the sacrament of the holy eucharist.

9. What are the chief benefits of holy communion?

(a) Union with Jesus Christ whom we really and truly receive.

(b) Many supernatural graces and blessings, e.g., the forgiveness of venial sins, the strength to resist future temptations.

(c) The pledge of everlasting life in heaven.[8]

HOLY ORDERS

1. What is the sacrament of Holy Orders? Holy Orders is the sacrament through which Jesus Christ bestows on certain members of the Church a permanent charism of the Holy Spirit for special service of the people of God.

2. Why is this sacrament called holy orders? Because it comprises three steps or grades: deacon, priest, and bishop.
3. Who can receive holy orders? Any Catholic of the male sex who has the necessary qualifications and is chosen by a bishop.
4. Who can administer the sacrament of holy orders? Only one who is at least a bishop can validly confer holy orders.[9]

MATRIMONY

1. What is the sacrament of matrimony? Matrimony is the sacrament in virtue of which a Christian man and woman signify and partake of the mystery of that unity and fruitful love which exists between Christ and the Church. They receive the graces necessary to discharge the duties of their state faithfully until death.
2. Was matrimony always a sacrament? No, before the coming of Christ into the world, matrimony, a sacred contract, was not a sacrament. Christ the Lord abundantly blessed this multi-faceted love of Christ through the sacrament of matrimony.
3. By whom was matrimony instituted? God himself is the author of matrimony. In the beginning he declared: "For this reason a man leaves his father and mother, clings to his wife, and the two become one flesh."
8. Can a sacramental marriage be broken? When the marriage contract is properly entered into between baptized persons, and the persons live together as husband and wife, the marriage cannot be broken except by the death of one of the parties. Jesus said: "Whoever puts away his wife and marries another commits adultery against her, and if the wife puts away her husband and marries another, she commits adultery." (Mark 10, 11-12).[10]

ANOINTING OF THE SICK

1. What is the sacrament of the anointing of the sick? The anointing of the sick is the sacrament for the seriously ill, the infirm and aged; by it the Church commends them to the suffering and glorified Lord, that he may lighten their suffering and save them.
2. How is this sacrament given? First, the priest prays over the sick person, and then he anoints the forehead and hands with oil made holy by God's blessing.
3. What does the priest say while anointing? The priest says: "Through this holy anointing, may the Lord in his love and mercy help you with the grace of the Holy Spirit. May the Lord who frees you from sin save you and raise you up.[11]

THE COMMANDMENTS OF GOD

1. Which are the principal commandments of God?--The Ten Commandments;

namely:

1. I am the Lord thy God, who brought thee out of the land of Egypt, out of the house of bondage. Thou shalt not have strange gods before me.
2. Thou shalt not take the name of the Lord thy God in vain.
3. Remember thou keep holy the Sabbath Day.
4. Honor thy father and thy mother.
5. Thou shalt not kill.
6. Thou shalt not commit adultery.
7. Thou shalt not steal.
8. Thou shalt not bear false witness against they neighbor.
9. Thou shalt not covet thy neighbor's wife.
10. Thou shalt not covet thy neighbor's goods.

2. Where do we find the Ten Commandments? In the Book of Exodus.[12]

What does an American Catholic Catechism have to do with mathematics? "Good and evil are categories."[13]

A CATHOLIC AMERICA--A MATHEMATICAL PERSPECTIVE

A Catholic Approach to Evolution and Spiritual Life
Father E.J. Cuskelly, M.S.C.
"Grace Perfects Nature"

Theology is the discipline of grace perfecting nature. It is the analysis of the evolution of the spiritual in psychology. It has been called the "study of Christian perfection." Father Cuskelly contends that man has three urges--"money, power, and sex." The Church has built up to counter these urges the disciplines of "poverty, chastity, and obedience." Theology tries to develop the Christian personality. The theology of grace, for example, can be approached in personal categories. That is, "sanctifying grace is a quality inherent in the substance of the soul; the infused virtues are proximate principles of operation; the gifts of the Holy Spirit are potencies; actual graces are helps of illumination and inspiration."[1]

God's gift is supernatural. It is outside the arena of experience. God brings us into a relationship of friendship and love with the Trinity. We are drawn into an intimacy with God. Grace is freely given. God created us to see the beatific vision. Man was created for divine life. The love of God is powerful. "Grace is enjoying the Divine." Truth is the holding on to the propositions set forth by the Church to be held on the authority of the almighty. Faith is the adherence to God and the acceptance of his route to salvation. God is spirit. We can approach him through love. God approaches us through Grace. God loves us first. The learning of things of God causes intimacy says St. Thomas Aquinas.[2]

God's love produces goodness in life. How does a person meet God as a friend? First, he makes himself a friend to us in a personal approach. Second, man accepts the personal approach of God in friendship with God.[3]

Man's Personality

Father Cuskelly contends that "a human person is an individual existing in a free human nature." Man can have a personal relationship with God. "Thou hast made us for thyself, O God, and our hearts are ever restless until they rest in thee." We have "a heart to know thee." Marriage is the sacrament of personal commitment. "The Church is the bride of Christ." Celibacy is another way to vow one's way to God. "Thou hast conquered, O Pale Galilean, the world has grown grey with thy breath." Self-denial is an important part of the spiritual life. We should deny ourselves to become enraptured in Christ's personality. The adult has extreme egoism. "Taking into account the needs of men today, the best way to cultivate the spirit of renunciation which the Gospels declare to be essential, seems to be to make religion more and more personal, to put more and more thought into it." We should be completely dedicated to God. The point of departure is Faith. We believe that it is "God who loves us." Humility is the terra firma of the spiritual existence. In man's personality the mystical should be identified with the personal.[4]

A CATHOLIC AMERICA--A MATHEMATICAL PERSPECTIVE

Prayer

Prayer is the expression of our love and obedience to God. We should love the presence "of God as Faith." We should want "God and his will." We should praise God. Faith is knowledge of God as our personal Savior. Prayer is a manifestation of friendliness with God. We can test our strivings with prayer by four purities: (1) "purity of conscience," (2) "purity of heart," (3) "purity of mind," and (4) "purity of action."[5]

Prayer and Progress

There are four stages of prayer: (1) meditation wherein reason predominates, (2) emotional prayer, (3) contemplation, and (4) the "night of the senses." We should pray to the Blessed Mother. Mary is the "mother of God." Mary is the "mother of men." Mary is the "Virgin Immaculate." A contemplative lives a full spiritual life. We should pray and act on prayer.[6]

Satan's Realm

Satan is "the Father of Lies." He betrays the individual's soul by it's particular vulnerable point. He seeks periods of distraction. He takes control of little things and the soul falls gravely. He hides vice under the "appearance of virtue." The "Father of Lies" acts through disceit and illusion, flights away from reality, and "scruples."[7]

Grace

God draws us deeper and deeper into the mystery that is Christ. There are several stages in the development of the Christian personality. There is the formative stage and the transition stage. Here we mature spiritually. There is the unclassified stage where some do not mature fully spiritually. And there is the stage of adult problems. Here there is a goal of further "purification" to free us from pride and willfulness.[8]

Are mathematical approaches to evolution valid? Maybe so. Think about the double helix.[9] It is a key to evolution.

A CATHOLIC AMERICA--A MATHEMATICAL PERSPECTIVE

C.S. Lewis: A Christian Apologist

People quarrel over standards of conduct. People try to show what is wrong. This law about values of Right and Wrong used to be designated as the "Law of Nature." This law is different from laws of physics and nature. It is called the "Law of Human Nature." For example, a persons's body cannot disobey the laws of heredity or physics. However, a person can disobey the "Law of Nature," if he or she wishes. Man thought that all men had knowledge of decent behavior. Did different civilizations have different moral teachings. Lewis contends that their moral teachings were much like ours. Running away in battle and selfishness have had few admirers. People break the "Law of Nature." They try to change rules and shift responsibility. Don't all people have a herd instinct? Lewis doesn't deny that there is some truth to the idea of the herd instinct. But he says that it is not the "Moral Law."[1]

Man feels two desires. First, is a desire to help a man if he is drowning. Second, is a "desire" to stay out of danger. There is also a third element that indicates to you that you should judge between these two predispositions. The "Moral Law" tells us which song to sing. Our instincts are only the keys.[2]

The "Moral Law" is not one of man's instincts. Man does not act from his instincts if he makes one instinct larger than another instinct. "The thing that tells you which note on the piano needs to be played louder cannot itself be that note." "Every single note is right at one time and wrong at another." The "Moral Law" directs our instincts such as sex and hunger.[3]

Is "Moral Nature" relative? No. It is similar to some truths found in mathematics. Is it a materialistic universe? Is it a spiritual universe with a mind behind it? It is difficult to answer these questions. For example, science explains how the universe is. Religion, in contrast, explains why. We know more about man than science can tell us. It seems that something is directing the universe. The universe is beautiful and frightening. A clue to the universe is the "Law of Nature."[4] "You cling to the Supernatural or you will find that the natural denigrates into the unnatural."[5]

Why should not a religion be complex if the world is not simple? Dualism is the belief that there are two separate powers behind everything. One source is good and the other force is bad. If one power is good and the other power is bad, then we need a third power that sets up standards of right and wrong.[6]

Satan wanted to be God. Thus our world is rebel occupied. We have free will. Man runs on God like a machine runs on electricity. God forgives sins. Christ underwent the perfect death, surrender, and humiliation. He calls us to baptism, belief, and Mass. Christ acts through us.[7]

Morality is focused on three things: (1) "fair play and harmony between people, (2) harmonizing a person's interior life, and (3) life's purpose." "There are seven virtues." There are four "Cardinal virtues" and three "theological virtues." The

26

A CATHOLIC AMERICA--A MATHEMATICAL PERSPECTIVE

Cardinal virtues are "prudence, temperance, justice, and fortitude." The "theological virtues" are faith, hope, and charity. Social morality is to love God and love our neighbor.[8]

Forgiveness is one part of charity. Love is a state of the will. Hope is a focus on the next world. Faith needs to be fed by prayers.[9]

"Beyond Personality"

Men will become perfect if they devote themselves to Christ. Lewis argues that there are "Personalities" in God and that there are no real personalities elsewhere. You will find in yourself only "hatred, loneliness, despair." If you look for Christ you will find him and all else that counts. You should try to transcend your own personality to that of Christ's personality.[10]

C.S. Lewis did not convert to Catholicism. However, his writings seem to be more orthodox than many Catholic writers. Personality has a mathematical dimension. For example, Cattel has written a book on the scientific approach to personality.[11]

A CATHOLIC AMERICA--A MATHEMATICAL PERSPECTIVE

Godel Numbers

Godel develops a systematic calculus that contains all the usual arithmetical symbols that can be set forth and that can develop familiar arithmetical relations. This calculus is built from a group of simple signs. Godel says that we can assign a special number to each of these signs. The number is called a "Godel number"of this proof. There are two kinds of these signs: "the constant signs and the variables." There are ten constant signs which Godel numbers from 1 to 10. And there are three kinds of variables that are included with the constant elementary signs. These variables are "x", "y", and "z." They can be used to expand numerals and numerical expressions.[1]

There are three or more variables "p", "q", and "r." They can be used to expand formulas or sentences. And there are predicate variables "P", "Q", and "R" for which a grammar such as "Prime" or "Greater than" can be represented. These variables are given Godel numbers which are developed by (1) a separate numerical variable--a orime number larger than ten; (2) with each sentential variable the square of a prime number larger than ten; and (3) with each separate predicate the cube of a prime number larger than ten.[2]

If a number is "less than or equal to ten, it is the Godel number of an elementary constant sign." If the number is larger than 10 it can be broken down into prime numbers. "If it is a prime number greater than 10 or the second or third power of such a prime, it is the Godel number or an identifiable variable" If the number is the "product of successive primes... it may be the Godel number either of a formula or of a sequence of formulas.[3]

Because each expression in the calculus is related to a Godel number, a meta-mathematical statement regarding expressions and their relations to each other can be expanded as a declaration about Godel numbers and their relations to each other. Every meta-mathematical expression is represented by a special arithmetical formula. There is then a logical relationship between meta-mathematical relations and their arithmetical formulas.[4]

The Bible: A Book of Numbers?

Is the Bible Godellian? Probably not. But many numbers that Godel designates as Godel numbers are in the Bible. For example, the number "one" occurs 2583 times. The number "two" occurs 847 times. The number "three" occurs 489 times. The number "four" occurs 315 times. The number "five" occurs 344 times. The number "six" occurs 202 times. The number "seven" occurs 453 times. The number "eight" occurs 81 times. The number "nine" occurs 491 times. The number "ten" occurs 224 times. The word "some" occurs 398 times. The word "all" occurs 5350 times. The word "every" occurs 971 times. The word "everything" occurs 136 times. The word "nothing" occurs 280 times. The word "nought" occurs 18

A CATHOLIC AMERICA--A MATHEMATICAL PERSPECTIVE

times. The word "zero" is not in the Bible. The prime number "eleven" occurs 23 times. The prime number "thirteen" occurs "13" times. The prime number "seventeen" occurs 9 times. The word "multiply" occurs 53 times. The word "divide" occurs 40 times. The word "add" occurs 31 times. The word "subtract" is not found in the Bible. The word "same" occurs 193 times. The word "inverse" is not found in the Bible. The word "identical" is not found in the Bible.[1]

A CATHOLIC AMERICA--A MATHEMATICAL PERSPECTIVE

Thesis: American Catholic Intellectual History is also an extension of mathematics. For example, it is based on a time line. Intellectual history is a category within history.

"I Will Draw All Things to Myself."
Mary Boggie quoting Jesus

Margaret Reher's American Catholic Intellectual History

The year 1789 is a crucial date in the history of America and in the history of the American Catholic religion. In 1789 George Washington was made president of the republic; furthermore, John Carroll was chosen as America's first Catholic bishop. There seems to be a coincidence between these two historical moves. When Washington assumed office, the Constitution was only an untested document. When Carroll started his office, the structure of the Catholic Church existed mainly in his mind.[1]

Because thoughts don't usually exist in a vacuum, notice should be given to the intellectual movements that touched upon the Church as it struggled to evolve its mission. The main intellectual movement that the Church had to come to terms with was Enlightenment thought. The term Enlightenment is hard to define. Dogma was its primary enemy. Out of Enlightenment thought came cries for freedom of conscience and cries for tolerance of different religions. John Locke(1632-1704) made a plan for the coming together of the diverse Christian theologies. However, he did not include the Catholic Church in his plea for union.[2]

The Enlightenment is usually argued to be Protestant. But, there was a clique of European Catholics who tried to unite the gulf between the Church and the thinkers of the Enlightenment. These thinkers resisted the idea of the Church which was for a union of Church and state; and, the idea that error had few rights. These writers influenced the thought of John Carroll and some other Catholics. Arthur O'Leary(1729-1802), Joseph Berington(1743-1827) and John Fletcher(1766-1845) were all from England and were recipients of the idea that the religion of the people was the religion of the king. They also had respect for constitutional governments. This tradition allowed them to work with the problem of religious pluralism without focusing on the problem of doctrinal apathy or doctrinal error.[3]

Fletcher, for example, differentiated between allowable and inadmissible forms of Catholic prejudice. He contended that the Church's rule should be in the spiritual area and that the Church had no jurisdiction over a person's life, liberty, and property. Fletcher supported the Church's opposition to mixing belief with error.[4]

O'Leary contended, also, that life, liberty, and property were impartial rights that shouldn't be changed by religion or by civil rulers. He maintained that conscience and religion were prior to a social contract. O'Leary fought against the idea set forth by John Calvin that religion should be controlled by the state.[5]

Moreover, Berington contended that no religion should have the backing of the state. Thus religious tolerance was a human right. State problems and conscience were separate. A good citizen only had to obey the regulations of the land. For a while, John Carroll was deeply influenced by Berington. Thus while Berington, Fletcher, and O'Leary were able to mold a united avenue to Enlightenment thought and to Catholicism. They were not able to see their ideas actualized.[6]

Charles Carroll, a signer of the Declaration of Independence, helped put this synthesis into reality. He took a position for separation of state from religion. He argued that "rational investigation is as open to Catholics, as to any other set of men." Carroll also set forth the idea that freedom of thought included his theological ideas in the statement of his talk on papal infallibility. He argued that the focus of infallibility resided with the Pope, in a "general council," or with the pope together with the Pope's council received by the Church. Carroll's councillar idea of infallibility and his idea of the particular infallibility of the Pope was an approach that lasted for a long time into the nineteenth century.[7]

Are Protestants going to be damned? Carroll suggested that we should believe that they could be saved. He saw that religious pluralism was a good.[8]

It was the problem of lay initiatives that created doubts in John Carroll's mind about the English Catholic, Joseph Berington. In his formative dealings with St. Peter's Church in New York, Carroll heard that Berington allowed the nomination by the parishioners of their pastors. Carroll felt that this approach would be harmful if tolerated in America.[9]

In Carroll's mind, the two greatest problems towards a reunification of different Christian groups were "the Latin tongue in the public liturgy," and the lack of precision in the jurisdiction of the Vatican. Carroll also opposed Berington's attack against the clerical vow of celibacy.[10]

Carroll took an interest in Elizabeth Ann Seton, a convert to Catholicism, and her five offspring. After a move from New York to Baltimore, Seton started a school. Later on, she became convinced that she had a vocation. During 1809, Bishop Carroll recognized the vows of the first sisterhood, the "Sisters of Charity of Saint Joseph." Later, Elizabeth Seton became, in 1975, the first saint in America.[11]

The Spiritual Exercises of Saint Ignatius were a critical factor in John Carroll's life. Another work that was formative in Catholics' lives was Saint Francis de Sales, An Introduction to the Devout Life. This book portrays a portrait of the spiritual development of a Christian leader. A third aid to spiritual change was membership in the Sodality of the Blessed Virgin Mary.[12]

The pious quality of the Anglo-American Catholics was similar to that of the English Catholics. John Gother, for example, a convert, developed several manuals of prayer. Gother's disciple was Richard Challory. His best known work, Garden of the Soul, became a classic. This book emphasized a personal relationship with

A CATHOLIC AMERICA--A MATHEMATICAL PERSPECTIVE

Jesus.[13]

The first Catholic book printed in America was a translation of Francois Fenelon's, Dissertation on Pure Love. It was published in 1738. The earliest Catholic book, The Manual of Prayers, was published in 1774. During 1784, Christopher Talbot, the first Catholic publisher in Philadelphia, reprinted Joseph Reeve's, New History of the Old and New Testament.[14]

John Carroll defended the American Church to discontented priests; and, he defended the Catholic Americans as loyal American citizens. He contended that freedom should be welcomed by both Protestants and Catholics. Carroll came out for the separation of Church and state when it was usually thought by many that a state religion was needed. John Carroll's brother, Daniel, helped shape the 1791 amendment to the Constitution that forbids Congress from making a state religion.[15]

During 1789, a papal bull, Ex Hac Apostolicae, was signed by Pope Pius VI. In that year, John Carroll was made the first Catholic bishop of the New Republic. Shortly thereafter, before John Carroll received episcopal consecration, the French revolution broke out. John Carroll's cousin, Charles, was enthusiastic when the titles of French nobility fell. However, he was shocked when the king, King Louis XVI, was guillotined during 1793. John Carroll was horrified by French terrorism.[16]

While French anticlericalism continued, John Carroll received an offer from the superior general of the Sulpician order, Mr. Jacques-Andre Emery. He told Carroll that the Sulpicians wanted to establish a seminary in the United States. When the newly elevated bishop came home in 1791, he found four Sulpician priests and five seminarians had already arrived in Baltimore. The stage was prepared for a difficult story involving Carroll and the Sulpicians.[17]

After Carroll returned to America, a project of his was printed by Matthew Carey's press during December 1, 1790. Earlier, Carey enlisted Carroll's support about printing an English version of the Douai-Rhiems bible. Carroll backed the plan. Carey collected almost twenty different copies of the Scriptures before he printed the project. Carey goes down in history as the publisher of the first Catholic American Bible.[18]

Earlier in his life, at age fifteen, Matthew Carey became an apprentice to a printer. Several years later he printed a treatise entitled, "The Urgent Necessity for a Repeal of the Penal Code against Roman Catholics." This treatise was so controversial that Carey had to flee to France where he worked under Benjamin Franklin. Later on Carey went to Ireland where he continued to speak out against the English Parliament. During 1784, he fled to the United States dressed as a woman.[19]

After his return, Matthew Carey started a career in writing and publishing that brought him recognition as Philadelphia's foremost Catholic intellectual. Carey helped make Philadelphia the book capital of America. He made his efforts strike gold with his, The American Museum. Started in 1787, The Museum was considered the best literary effort of its type in the United States. George Washington praised this literary effort as a very useful enterprise. Carey was not, as it is noted, a Catholic publisher,but, a publisher who was a Roman Catholic.[20]

Both Charles and John Carroll were influenced by the Enlightenment. They accepted America's history and the place of the Catholic Church within this history. A friendly Protestant historian wrote that John Carroll's "argument for a disestablished and non-imperial faith inside a pluralist republic is his special legacy."[21]

John England was the bishop who adopted the Enlightenment thought of the Carrolls. He was given the see of Charleston, South Carolina in 1820. John England was faced with the problems of "trusteeism and nativism." He maintained that the greatest problem faced by the Catholics was a failure to keep Church and state separate. One of England's earliest efforts was his attempt to prove that Catholicism was compatible with republicanism. The first issue of England's, America, journal which was dedicated to the defense of Catholic doctrines was printed on July 5, 1822. England pointed out the five most important elements of Protestant prejudice: they were as follows: "family loyalties; respect for teachers who instill anti-Roman sentiment; flagrant misuses of history, of the sciences, and belle lettres."[22]

England, like Carroll, tried to defend the idea that Catholicism was compatible with republicanism. He pointed out, for example, the government of Catholic monastic orders. Other examples were the various Catholic political entities such as Genoa, Venice, San Marino and the Swiss cantons. England was for making a differentiation between the Church's spiritual reality and America's political constitution. He argued that "God and Christ" are the highest legislators and the Pope was a religious presider, not an earthly monarch. England contended that the Church is like the supreme court. Church dogma is like different laws the court adopts; the papal legate is like an American ambassador; the Church's congress is the council, and the bishops are like congressmen.[23]

England did more than throw a bouquet towards the idea of constitutional government. He drew up The Constitution of the Roman Catholic Church of North Carolina, South Carolina, and Georgia. This Constitution outlined the obligations and rights of all the Church members. It established a voluntary association for Church members and created a corporation for dealing with ecclesiastical funds. It did not, however, deal adequately with the position of women in the Church. According to England, the highest court of the Church was a general council which worked together with the pope. England's Constitution set up district churches which were drawn up by Church membership. Each church was ruled by a vestry containing clergymen and laymen. Different vestries could outline their own form of government within constitutional rules. However, England pointed out that a bishop had the right to appoint pastors or remove them.[24]

England drew up a second constitution for the Philosophical and Classical Seminary of Charleston in 1822. This seminary was founded on the ideas of mutual aid and charity. It restricted the power of the president. Moreover, England started a school for black children. Later on, he defended slavery partly because there was among the abolitionists a strong anti-Catholic prejudice.[25]

England continued defending the American approach to government after Pope Gregory XVI wrote Mirari Vos during 1832. The Pope favored union of church and state; and, he spoke out against liberty of conscience as an act that would promote religious apathy.[26]

Both Carroll and England differentiated between religious freedom and religious indifference. At the time of the first Provincial Council of Baltimore interfaith friendship had taken a down turn. Irish, French, and German Catholics posed a threat to Protestant dominance. Nativism broke out periodically. The Church's response to nativism was separatism.[27]

The separation between Church and state that John Carroll and John England advocated was accepted by many people in the Catholic community. The American bishops failed to lead their communities on political issues. They were silent on the social justice issues of slavery and the removal of the Indians from the South.[28]

Two themes from the twentieth century religious history that could be applied here are solidarity and renew.[29] The early Catholic American community worked as partners with a lot of solidarity for the American revolution and with the American Protestants. The Enlightenment bought on a renewal of the Catholic religion in America. Catholics turned over a new leaf in the New World.

Baltimore's First Provincial Council was ended at the close of the decade that was beginning to show America's changeover to an industrial civilization. Many people were joining the factory and labor group. America's population was becoming much more urbanized. Along with the national expansion, urbanization and industrialization came floods of new immigrants. Large numbers of Lutheran Germans, Scots-Irish Presbyterians, and Catholics began to fill the country. The Roman Catholics were the largest number of immigrants and they came from a very diverse group of nationalities. During the 1850's, for example, Catholic immigrants were almost a million people.[30]

The job of the Catholic Church was complicated. First, to fend off nativist attacks, the values of these Catholics had to be spelled out to American Yankees. Second, the American way of life had to be set forth in the minds of the new immigrants. Third, while these new immigrants were loyal to the Catholic Church they had a "romantic" intellectual bent. What was "romanticism?" It was a movement against the rationalism and objectivity of the Enlightenment.[31]

Johann Mohler was a dominant leader in the Tubingen school of thought that tried to fight the rationalism of Enlightenment thought. He developed the German idealist school of thought. The publication of Mohler's Symbolik in 1843 was a "theological event." He was a major figure in developing the historical and intuitive approaches towards the analysis of doctrine. Another Catholic figure was Count Joseph de Maistre(1754-1821). His book, Du Pope, was a defense of the Papacy and infallibility. In it he was searching for a principle of togetherness to try to make sense of the French Revolution. He argued that the Catholic synthesis made during the

A CATHOLIC AMERICA--A MATHEMATICAL PERSPECTIVE

Middle Ages was true for every period.[32]

Romanticism wasn't either completely European or Catholic. The American Protestants had some movements towards unity. The "Mercersburg movement" was one of these movements; it was led by Philip Schaff and John W. Nevin in the German Reformed Church. The Transcendentalists and the Episcopalians who were influenced by the Oxford movement were also influenced by Romanticism. It has been suggested that Roman Catholics moved more slowly than either the Europeans or the Protestants to assert Romanticism. The Catholics relied more on reason. Their view was developed by such leaders as Orestes Brownson, Isaac Hecker, and James A. McMaster.[33]

The Enlightenment school of thought was not completely forgotten. Father Isaac Hecker was a "liberal" follower of the Enlightenment. He asked the Church to assent to reason, the idea of progress, and the idea of liberty. Other more "'conservative"' romantic thinkers like Orestes Brownson tried to restore a single religion, the Catholic Church, to solidify the social aspect of America. Brownson made significant contributions to the different religions and intellectuals of his time. He was delivered in Stockbridge, Vermont during 1802. His father died a short time after Orestes' birth. His mother sent, because of poverty, Orestes to live with an older couple. This couple didn't practice a religion.[34]

Possibly because he lost his father and was raised without a creed, he developed a strong interest in both fatherhood and in a Christian creed. During his teens, Brownson, lived with his mother and received a brief education. In his nineteenth year he became a Presbyterian. However, Brownson found the Presbyterian ideas of predestination and eternal punishment hard to believe in. He soon became a Universalist because he believed in the salvation of all. During 1826 Brownson became a Universalist preacher. In 1829 Orestes became the editor of a Universalist paper, Gospel Advocate and Impartial Investigator. He also started as a corresponding editor of the Free Inquirer. In this capacity he met Robert Owens and Fanny Wright social reformers. This group of reformers favored ending the "enemies of happiness": the Church, the family, and the right to private property.[35]

Brownson became a member of the Working Man's Party and argued for the idea of universal education of children. After a short time Brownson quit the Working Man's Party because he couldn't be a "party man." However, he never abandoned the cause of the poor. During 1830 Brownson quit Universalism. He said about this period that he had "rejected heaven for earth, God for man, eternity for time." Yet, by 1832 he became a Unitarian minister. He contended that only through religion could a belief in God, moral accountability and duty be found. The Unitarian religion interested him because leaders like William E. Channing were educated gentlemen.[36]

Brownson started an analysis of different rationalists, among them were the works of Benjamin Constant. Like Constant, Brownson saw religion as the sentiments of man developed in institutions; as civilizations aged they were criticized. Civilizations went through times of destruction and transition. These changes did not indicate the decline of society but progress. Religion progressed and so did man. It seems that the notions of the perfectibility of man, the importance of democracy, and the Christian tradition worked well together with the idea of progress.[37]

Brownson tried to develop a "Church of the Future" based on principles of progress. During 1836 he started his "Society for Christian Union and Progress." This "Church" was ecumenical. Brownson was not trying to start a new sect. He wanted unity and Catholicity as its goal. During this era, Brownson printed his New Views of Christianity, Society, and the Church. This book was heavily influenced by the French "utopian socialist," the Count de Saint-Simon.[38]

Looking back to Christian beginnings, Brownson saw Jesus as the God-man combining the antitheses of matter and spirit. Catholicism was the triumph of the spirit over the material. In contrast, Protestantism was more or less the triumph of the material over the spirit. Protestantism, for example, developed the state, liberty, reason, industry, and worldly pursuits.[39]

Authors who came after Brownson's New Views was published saw some of the pitfalls of the Enlightenment. They didn't agree with the individualists' interpretation of the American and French revolutions. Brownson had a romantic approach to history. In 1836 Brownson started another journal, The Boston Quarterly Review. Its purpose was to use literature, religion, and philosophy to defend democracy and democratic reform movements.[40]

During the twenties and thirties when many Working Men's Parties were created after the depression of 1837, Brownson wrote an essay "The Laboring Class." He spoke out against industrialism. In his point of view he thought that no two Christian countries were as bad to the poor as the United States and England. Corporations, Brownson maintained, lack a soul. He argued that government should control its own powers, destroy monopolies, and radically change the banking system. He also was opposed to the system of inheritance laws.[41]

The political campaign during 1840 was very difficult. A former governor of Georgia, John Forsyth, argued that William Harrison was backed by Catholics opposed to Masonry and by Protestants who were for ending slavery. Forsyth also argued that the English abolitionist society had influenced Pope Gregory XVI who condemned slave trading. John England answered this suggestion by raising the question of where the English influence was with the Popes from Pius II to Pius VII who had made similar statements.[42]

During 1841, Brownson met Isaac Hecker who was also working in labor politics. Isaac Hecker was fathered in 1819 of German immigrants. Like Browson, Hecker was aligned with Jacksonian democracy. He also aligned himself with the Loco-Focos, a group that was opposed to monopoly.[43]

Isaac Hecker went to a lecture given by Brownson called "The Democracy of Christ." The theme was that "Christ was the Big Democracy and the Gospel was the true Democratic platform." Brownson became Hecker's intellectual and spiritual

A CATHOLIC AMERICA--A MATHEMATICAL PERSPECTIVE

mentor.[44]

In the year 1842, Brownson participated in the Transcendentalist experiment, Brook farm, in Massachusetts. It was a utopian community led in part by Ralph Waldo Emerson. The Transcendentalists favored "divine immanence, intuitive perception of truth, and complete self-reliance." While Brownson was at Brook Farm he wrote an essay "The Mediationed Life of Jesus." This essay opposed William Ellery Channing's idea of the "divinity of humanity." Brownson argued for the idea of the depravity of humanity. He noted that little children were sometimes mischievous.[45]

Jesus was for Brownson a "providential man." He argued that God cannot be reduced to nature's laws. Brownson also thought that the Catholic idea of grace might be united with no damage or problems to the idea of reason in nature. He contended that Jesus was the mediator of grace. Brownson felt that we should live our lives in communion with God.[46] He said that "...**I have the conviction that I can be all the better Catholic because I am an American; and all the better American because I am a Catholic...**"[47]

During the decade of the 1840's both Brownson and Hecker watched the English and Oxford movements develop. John Henry Newman had earlier set forth an argument on the merits of apostolic succession. In America the idea of Church union came before the Oxford movement was safeguarded by the High Church. In Anglican England, there was hostility towards the Tractarians.[48]

Francis Patrick Kenrick, a Philadelphia bishop, was an ardent follower of the Oxford movement. In 1838 he invited the Protestant Episcopal bishops to move towards Catholicism. He argued that these bishops should read John Milner's, A English Catholic Bishop--The End of Religious Controversy. This book was a standard book of Apologetics. Henry Hopkins, a Vermont Episcopal bishop, countered by saying that the church of Rome had corrupted the early purity of the Church. Kenrick's main contribution is his emphasis on episcopal collegiality. However, while Kenrick was a collegialist, he was not a conciliarist. Should the ecumenical council overrule the Pope? He replied that the council should have a head; it shouldn't be acephalous. Kenrick was one of the first to translate Roman law into Anglo-American common law.[49]

Isaac Hecker interviewed both the Episcopal and Roman Churches about dedicating his life to Church work. He decided after several interviews to become a Catholic. Hecker was baptized during August 1844. Brownson converted several months later. And in the following year John Henry Newman, one of the main leaders of the British Oxford movement, did also. Brownson contended that a strong and conservative Church could save America. The Catholic Church was his answer.[50]

According to Brownson, America was a divided country. He felt that the American constitution and the English order were at loggerheads with the new Jacksonian democracy. The Jacksonians were heirs to the Anti-Federalists, French Revolutionists and anarchists. Brownson contended that "constitutional republicans" worked well with Catholicism.[51]

Brownson worried about the efficacy of "democracy" because there were a growing number of proletariats which were spawned by the industrial revolution. He argued that the Pope should be able to depose heads of state. An outcome of this position was that Brownson was harried by both Catholics and non-Catholics. Francis Kenrick said, for example, that Brownson should be more "temperate." Archbishop John B. Purcell called for Brownson's excommunication.[52]

Looking backwards, Brownson's son, Henry, argued that his father's position on the temporal powers of the Pope was "impudent and likely to expose the church to unnecessary odium." Brownson, however, didn't change his position.[53]

James A. McMaster, editor of the Freeman's Journal, was like Brownson and ultramontane. Both of these thinkers thought that papal primacy should include the right to "depose rulers and absolve subjects of their allegiance."[54]

While Brownson's relationship to Rome was creating problems for him at home, Hecker was having problems due to an excessive attachment to America. In the year 1855 Hecker published an almost autobiographical book call Questions of the Soul. In this book, Hecker tried to demonstrate that "the destiny of the soul is to be one with God...and without God man is incomplete and his actions are ineffective." He maintained that Protestant emphasis on individualism caused much of the apostasy of the era. Humanity and science were displacing God and Judaeo-Christian values. Hecker took a new approach to Catholic literature. Instead of a dogmatic defense of the Catholic religion or an attempt to argue against the errors of Protestantism, he gave an expose of the depths of man's spiritual nature.[55]

Brownson's book, like Hecker's, Questions of the Soul, was a breath of fresh air. The book was written to show that the Catholic Church could best meet the needs of the heart. To answer the idea that the Catholic Church meets the desires of the mind as well, Hecker wrote Aspirations of Nature to meet this need. Like his earlier book, Questions, Hecker argued for the goodness of man's nature. Because intelligence is somewhat limited, revelation is demanded. The Catholic Church, Hecker maintained, can satisfy man's spiritual needs the best. This idea became the fountainhead of Hecker's later theological writings.[56]

While he was in Rome, Hecker wrote a treatise for the Jesuit periodical, Civilta Cattolica. He tried to explain in this article the Church's position in America. Hecker argued that nations like individuals are given the grace of a convert. He drew upon his experiences like Brook farm; and, Hecker contrasted these experiences with the European varieties of socialism. The European socialists repudiated the Church of Rome, while the American varieties moved toward conversion.[57]

God helped direct the Founding Fathers and through God the Christian faith would prevail. Americans were able to self-govern themselves. American democracy argued for natural law and rights and defended the idea of human nature and justice. This is what the Church also teaches.[58]

To counter the argument that Americans are materialistic, Hecker claimed that

A CATHOLIC AMERICA--A MATHEMATICAL PERSPECTIVE

American money worked towards its conversion. Americans did not have to fear control by their bosses because there was a need for labor. When the Pope brought up to Hecker America's materialistic pursuits, Hecker said, "true, Holy Father, but the holy faith is there."[59]

The pope dispensed Hecker and several other priests from their vows. He encouraged them to develop a community to work towards America's conversion. During 1858 Hecker and several others formed the Missionary Society of Saint Paul the Apostle.[60]

Hecker thought that religion was pertinent to all areas of life, including the political arena. He also thought that America's theory of natural rights was similar to the Catholic idea of natural law. Hecker argued that when America converted to the Catholic religion, America would be a repository of virtue.[61]

Brownson, in contrast, wondered if democracy and Catholicism were compatible. Hecker thought that the American experiment would be part of the Catholic idea of liberal politics. At the time that Hecker developed the Paulists, he tried to build a similar group for lay women. This attempt failed. Hecker maintained that the Church was a positive factor in the women's movement; Hecker heralded the rights of women.[62]

The need to define the idea of ecclesiastical authority in the age of various kinds of revolution led Pius IX to convene the First Vatican Council in the year, 1870. It became apparent that there was a firm faction growing that favored papal infallibility. Most of the American bishops opposed the move because they thought that the doctrine of infallibility would inflame nativist sentiment. On July 18, 1870 a vote was cast on the new Constitution, <u>Pastor Aeternus</u>. Papal infallibility and legal primacy won out. After a time, all the American bishops favored the new definitions.[63]

Brownson thought that "papal primacy" should include the right to depose political leaders. He felt that the temporal power of the pope should be recognized. Protestants in America saw this document as a threat to the New World's ideas of "freedom and democracy." Napoleon, Garibaldi, and Cavour persecuted the pope until Pius IX was a "prisoner of the Vatican." Brownson died in 1874.[64]

Hecker didn't separate the concept of individualism from the idea of sanctification. He had a very positive conception of the laity. Hecker argued that the Church is a democratic institution. He contended that the beginning and destiny of America and the Catholic Church were one and the same. Hecker didn't accept the legitimacy of the idea of monarchy on theological grounds. Hecker has been called "the Yankee Paul."[65]

"The Catholic University"

The notion of developing a Catholic University in America was first talked up by Archbishop Martin John Spalding while he was preparing for the Second Plenary

A CATHOLIC AMERICA--A MATHEMATICAL PERSPECTIVE

Council of Baltimore. The bishops met this idea without much enthusiasm. The problem of developing this project was left to John Lancaster Spalding, the archbishop's nephew. At the same time, during the leadership of Leo XIII, neo-Thomism made a comeback. Pope Leo XIII's encyclical, Aeterni Patris, defended scholasticism and decreed that Thomism was to be the philosophy taught in all Catholic institutions of higher learning.[66]

In an essay commemorating the hundredth anniversary of the Declaration of Independence, Spalding painted an arid picture of America's intellectual contributors. The solution that Spalding offered to this dilemma was a university where a Catholic elite would be trained. In the year, 1878, Spalding was offered the opportunity to promote the idea of a Catholic university. The cornerstone for a Catholic university was laid in 1888. In his talk at the opening of the university, Spalding played up the works of the United States in promoting "fraternity and equality." He contended that the university should be progressive. His address concluded by mentioning the name of the universities' benefactress, "Mary Gwendolyn Caldwell."[67]

During 1888 John Keane was made the first head of the Catholic University of America. He was influenced by Isaac Hecker, a friend. Keane was devoted to the Holy Spirit. He worked to converge American Civilization and the Catholic Church.[68]

In the 1880's Spalding promoted the education of women generally, and specifically teacher education for religious women. He advocated the establishment of a teacher's college. His key to knowledge was the development of a philosophic bent of mind. Spalding argued that women were sometimes superior to men in scholastic ability. An encyclical, Longinqua Oceani, was the first document that dealt with the American Church. It's praise of the Catholic University was "restrained."[69]

Evolution

To counter Darwin's Origin of Species which was published in 1857, George Mivart, a convert, allowed the hypothesis of evolution from lower life but proposed that the soul was created by God. John Zahm, a physics professor at Notre Dame, contended that evolution was consonant with scripture and with scholasticism.[70] However, it has been argued quite forcefully that evolution is a materialist fallacy.

Relativism and Dogma

The Americanizers favored the ideas of "progress" and "manifest destiny." Their hopes won out in several documents of Vatican II. Dignitatis Humanae Personae expressed the idea of religious freedom. Another document, Lumen Gentium, emphasized the laity and it moved away from a hierarchy toward more "episcopal collegiality." "The use of the vernacular" was one of the products of liturgical change set forth in Sacrosanctum Concilium.[71]

Spalding said that "all truth is orthodox." He argued for "free inquiry."

A CATHOLIC AMERICA--A MATHEMATICAL PERSPECTIVE

Spalding was a very gifted intellectual leader. He voiced the goals of the Catholic University and the Americanist goal of the Church in America.[72]

Modernism and Progressive Thought

What is modernism? Many priests rejected scholastic metaphysics and utilized historical methodology. Because the trends of Catholic modernist thought hit close to home, Pius X saw them as deadly sources of atheistic rationalism and "Protestant subjectivism." During September 1907 he issued Pascendi Dominici Gregis a systematic offense against modernism which he argued consisted as a "synthesis of all heresies."[73]

Did Christ have unlimited knowledge? Pope Saint Gregory maintained that Christ had unlimited knowledge. In contrast, Edward J. Hanna cited different scriptural lines--"Luke 2:52 and Mark 13:32" that seemed to place limits on Christ's knowledge. One of these quotes refers to the finding of Christ in the temple. It said that Christ grew in wisdom and knowledge. Rome got Hanna to write a follow up to his essay, "The Human Knowledge of Christ." Hanna then argued for the unlimited knowledge of Christ.[74]

The emphasis towards the centralization of Rome and towards ultramontanism wasn't changed until the American leadership rediscovered collegiality during Vatican II. Only a small number of priests quit the Church over modernism. One priest, for example, Joseph Slattery quit because he felt that the Catholic Church's relationship with black Catholics was disillusioning. He tendered his withdrawal in 1906. One exception to ultramontanism was the creation of the Catholic Encyclopedia. It was one of the great intellectual efforts of the Progressive era.[75]

Godel and American Catholic Intellectual History

Godel points out the limitations of the axiomatic method. What are some of the limitations of Catholic thought? Proofs of the existence of God have several contradictions. For example, if God is all powerful can he kill himself? Can God create a stone that he can't lift? Franey thinks that the answer to both of these questions is yes. He argues that "God can deal with contradictions that is why he is God."[76]

John Ryan

John Ryan was influenced by Hobson who argued that "underconsumption and oversavings are the main causes of industrial slumps and depressions." Similar to the Americanists that preceded him, Ryan, tried to reconcile the church with American Civilization. Ryan's doctoral dissertation, A Living Wage, was the earliest attempt in English to depict the Church's system of politics and economics. His thesis was that man's dignity is holy. He was influenced by Leo XIII's Rerum Novarum. Instead of condemning Marxism, Ryan tried to find the element of truth that exists in socialism.[77]

Locke had emphasized self-interest and the negative aspect of government. The Progressives felt that rights should be harmonized with responsibilities; and, they emphasized the positive role of government. Like Spalding who emphasized priestly culture, Ryan started a movement to make the clergy aware of the implications of economic issues. Examples of these issues are trusts, trade unionism, and profits and wages. Ryan promoted the idea of a living wage.[78]

In his most well-known book, Distributive Justice, Ryan argued against materialism; he contended that the best guide to social justice was religion. After WWI's armistice was completed on November 11, 1918, a Catholic "Bishops Plan for Social Reconstruction" was set forth. Ryan recast some of his ideas from his earlier works such as the idea of a "United States Employment Service" and the "National War Labor Board." Ryan also argued for the minimum wage, partnership between labor and management, cooperatives, and the end to child labor.[79]

In postwar America many Catholics accepted Progressive thought. However, new differences between Catholics and Protestants developed over such issues as prohibition, the Klan, and Alfred Smith's presidential attempt. A structure that survived the war was the National Catholic Welfare Council. During 1924 Ryan backed an amendment to the Constitution that would regulate child labor. It was never enacted due to lack of support. John Ryan's thought was based on neo-Thomism and thus differed from Modernism.[80]

Many Catholics felt that the state should recognize the Church as the true religion. The defeat of Alfred Smith was devastating for many American Catholics. John Ryan endorsed the New Deal. In contrast, Father Coughlin opposed Roosevelt. Coughlin argued that the country's monetary problems were a product of an elite of international bankers. The Spanish Civil War divided the Catholic community.[81]

The Movement Towards Pluralism: 1920-1968

During 1920 Neo-Thomism continued to exert a strong philosophical force. The magazine, The Commonweal, was started in 1924. It advocated a Catholic way of life. Alfred Smith, a Catholic, was defeated by Herbert Hoover. In the 1930's Father Coughlin argued that Roosevelt was a communist. Also, during the 1940's J.F. Powers and Flannery O'Conner started their writing careers. In the late 1950's Senator Joseph McCarthy attacked the State Department for harboring a number of communists. William Buckley started his conservative National Review. In 1960 J.F. Kennedy was the first Catholic elected president; and, Pope John XXIII started the Ecumenical Council. President Johnson signed a Civil Rights Bill into effect. The war in Viet Nam was also started in the 1960's. That war brings us to May 17, 1968 and the Berrigan's act of Civil Disobedience. Was this act a new beginning? Let us remember Thoreau and the Berrigans! The matrix of Catholic thought continues.[82]

American Intellectual History and Mathematics and Science

Why does not the Church hierarchy present a more mathematical analysis of the Church's theology and behavior? Why do not they have an ecumenical council on science and religion? Three avenues of this approach might be fissionism, fusionism, and field theory. Fissionism means a splitting apart, a division. An example of fissionism might be the splitting apart of the Marxist synthesis. Fusionism means a coming together to release tremendous energy. The world's Catholics working together are an example of fusionism. Field theory is the idea that within the space in the closeness of a particle, a field containing a form of energy exists. This particular field interacts with other fields and particles. In a spiritual way the Church acts like a field. It is a finite field with a large number of members. Margaret Reher's book, Catholic Intellectual History in America, and Thomas Reese's, Archbishop, describe the Catholic field; but, they don't come to terms with the relationship of Catholicism and the philosophy of science or with Godel's Proof and mathematics. For example, a new key to the Archbishops Pastoral on the American Economy is mathematics--math modeling of economic problems and behavior. The descriptive key to fissionism and fusionism is mathematics. The key to Catholic American intellectual history is to logically codify it.[83]

Catholic Intellectual History and the Future Research

Where specifically are the fields of Catholic thought and Catholic intellectual history going? It seems that the fields of Catholic thought and Catholic intellectual history are in "bad shape" compared to fields like economics and/or psychology. What can be done to rectify this situation? One answer seems to be to apply behavioral statistics to these fields. For example, indices of solidarity can be generated. What are some other examples of this approach? First, "descriptive statistics with frequency distributions and graphics, percentiles, measures of central tendency, and measures of dispersion, standard deviation, correlation, regression and prediction can be applied to verbal behavior and group behavior." Second, "inferential statistics with parametric tests of significance can be applied to independent samples." Third, "inferential statistics with nonparametric tests of significance" can be applied to different Catholic categories and scales. Parametric refers to a constant with a variety of values which is used as a "referent for determining other variables." This paragraph presents a thumbnail sketch of behavioral statistics; but, it seems to be a very fruitful approach to Catholic thought and behavior like voting behavior.[84]
Maybe Godel would find an undecidable proposition in behavioral statistics!

A CATHOLIC AMERICA--A MATHEMATICAL PERSPECTIVE

Abortion

Each man is a message. God creates individuals with a special song or message to send. No one else can deliver this message. It is entrusted only to oneself. Pro-life is the way to send the gift that is every person.[1]

In 1776 John Adams paced in tension with the knowledge that America's dream could die. He was considering three questions: "Is anyone there?" "Does anyone see what I see?" "Does anyone care?" Today we are living in the evening of the lack of respect for life. Father Powell focuses on these very questions: "Is anyone there?" "Does anyone see what I do?" Does anyone care?" He makes love the purpose of life. Love moves towards beauty not towards evil or bitterness. The Pro-life movement is a deed of love. Powell believes that a single person who loves is a powerful majority.[2]

The most moving incidents in John Powell's life were watching the birth of a child and his trip to Dachau. Today we are watching "the Nazi Nightmare in the American Dream." In Germany earlier there was extermination of the chronically ill. This extermination occurred at death centers. Hitler during September 1, 1939 had all the clients in state institutions evaluated by a group of doctors. These doctors spent approximately one week classifying clients. Most of these patients were stamped "Death." The evaluation forms became death warrants for the ill. The committee that handled these clients was called the "Realm's Work Committee of Institutions for Cure and Care."[3]

Another organization was built that was designed to handle the deaths of children. It was called the "Realms Committee for Scientific Approach to Severe Illness Do to Heredity and Constitution." There was developed a "Charitable Transport Company for the Sick" that brought clients to the Death Works. According to records 275,000 people were murdered. Powell calls the science of death "ktenology."[4]

Medical Research

Some of the Jewish people had experiments performed on them. For example, amputations and gun shot wounds were inflicted to test blood coagulants. During their confinement at Dachau three hundred clients were forced into freezing water. More than eighty people were killed in this manner. Other clients had injections of viruses and bacteria given them.[5]

In Holland when the Army Commander for the occupied Netherlands decided to rehabilitate patients for forced labor the Dutch physicians refused. Because of this not a single murder or sterilization was recommended by any Dutch Doctor. Today murder is still being carried out or supported by the "abortion, infanticide, and euthanasia movements." Today Doctors are creating a "guilt-free" civilization.[6]

A CATHOLIC AMERICA--A MATHEMATICAL PERSPECTIVE

America's "Silent Holocaust"

"Forty Years after Hitler: Yes, Again!"

On January 22,1973 the Supreme Court legalized abortion and started a new "slaughter of the innocents." Father Powell raises the point: Can a Christian be an innocent bystander at the mass killings?[7]

American Plight: "Death-on-Demand"

The case to be made here is that babies should, according to Dr Francis Crick, not be declared human unless they pass certain tests which if they should fail to pass certain babies would not be allowed to live. Dr. William Gaylin, a psychiatrist, told a group of university women that now the best thing for our children might be deciding which children to kill. Also, the best thing for our grandparents might mean that we have to kill them. An English physician, John Goundry, argues that "a death pill will be available...and used by force at the end of this century." This is an argument for a "New Ethic" that would take the place of the old morality.[8]

"Secular Humanism"

The rejection of religious ideas and values has also helped the "death-on-demand movement." Children are taught that we are beyond right and wrong, that "God is dead," and that man is God. This approach was first set forth in 1933 when John Dewey, the philosopher, and his friends wrote the "Humanist Manifesto." This manifesto stated that there isn't a God and that religious belief is an impediment to progress. Only science and reason can evolve us. The ten commandments are irrelevant.[9]

Patrick's Theses

What is needed today is a "Catholic anti-defamation league." Christianity and science should go together. We should head a call to holiness and become a Catholic America-a Papist America with Christian political parties.[10]

Is there an alternative to the "Nazification of the American dream?" Yes, there is. It is called the Just Life position. The Just Life people are opposed to abortion; but, they are for social and economic justice and they are for an end to the nuclear arms race. They have a consistent position on life. One could also be opposed to euthanasia, for animal rights, and for a Christian femininism. Each fetus, each life, is also a formula.[11]

A CATHOLIC AMERICA--A MATHEMATICAL PERSPECTIVE

"Archbishop"

Some Asides:

Can Godel Numbers be found in the Church or in the Bible? The answer is, yes.
Does the Bible have a Godel number or numbers? The answer is that there is an overlap of numbers.
Is the power structure of the Church, Byzantine. In part, yes it is.
Is the Church filled with Machiavellian intrigue? The account that Reese creates in his book doesn't support this idea.
Are the Clergy "White Mafia"?[1] The answer is not typically.

Thesis: The "underlying reality" of behavior is mathematics.
Thesis: Heaven is "Christian anarchy."[2] There are no masters or slaves. Government is the result of original sin.
The keys to the book, Archbishop, are solidarity, fusionism, concord, cooperation, fellowship, harmony, unity, fairness, reliability, soundness, stability, and mathematics.

"In our western civilization only one formal organization, the Roman Catholic Church, claims a substantial age."[1]

"The Selection of Bishops"

Mathias was selected to replace Judas as the twelfth apostle. In America, for contrast, bishops and archbishops have been selected by the Pope with the assistance of his advisor in Washington who has been named the pro-nuncio. It is difficult to become an archbishop. By canon law an archbishop must be single and Catholic, thirty-five years old or older and a priest for five years. Usually an archbishop is first a bishop and the leader of a diocese.[2]

Bishops and archbishops are chosen by the Pope. The candidate must be "a good pastor of souls and teacher of the faith." The Vatican is opposed to "politicking and pressure group activity."[3]

When different bishops have reviewed the candidates, they vote on them; then, the archbishop sends the selection to the pro-nuncio who resides in Washington, D.C. The report sets forth the candidate's "name, diocese, parents' names, schools attended and degrees received, date and place of ordination, foreign languages known...and appointments since ordination."[4]

The Papal Nuncio

The present Nuncio is Pio Laghi who assumed the position in 1984. When he receives the names from different provinces he organizes them alphabetically and by province and ethnic background. Some thinkers perceive that a priest who has been educated in Rome has a better chance with the congregation. Nearly fifty per cent of the archbishops have had some type of education in Rome.[5]

"The Pope"

The last step in the process of appointing a new bishop is taken when the prefect sets forth the pro-nuncio's, the congregation's, and the perfects own views which are presented to the Pope at a Papal audience. When the holy Father decides who the bishop will be, the congregation tells the pro-nuncio of his decision. The pro-nuncio then asks the priest, that is, the candidate, if he will take on the appointment. This approach takes from four to eight months from the time that there is an opening until the position is proclaimed.[6]

The Denouement

A number of variables become evident from a specification of the process through which a bishop obtains an office. The pro-nuncio is the significant person in the whole process. In promulgating his position in writing, the ternus, the pro-nuncio is capable of much influence in his paper, which accompanies the ternus. This report is written in Italian, the language of the congregation for Bishops. This paper is the primary report. Thus the pro-nuncio has a critical position in the choice of American bishops. However, the American bishops have an important role also. First, the bishops select priests in the provinces through which the bishops are selected. Second, by providing the ternus for the auxiliaries, the diocesan bishops have a critical power on the American hierarchy.[7]

A third variable in the process is the diocesan bishop who is near retirement age. He initiates the process and needs of his diocese and elaborates the qualities that the bishop who follows him should have. He suggests to the pro-nuncio the different names of persons who will help advise the pro-nuncio about the particular diocese and the next bishops appointment.[8]

The fourth group of actors in this selection process are the different priests who fill in the questionnaires on the different priest candidates. These are the primary diocesan leaders listed in the Official Catholic Directory. The fifth position is the role of the Congregation of Bishops and their prefect. Their job seems to be largely a check on how the pro-nuncio fulfills his role. The sixth variable is the position of the Pope which is very important. He can select as a bishop any priest he desires.[9]

Thus the choice of bishop is not a democratic one but an institutional selection that tries through large consultations to find a candidate who would become a good pastoral bishop, who is concerned about the well-being of the different people in his particular diocese, and who is loyal to Rome and to the Pope. It is an autocratic method of selection. Besides the good will of the participants, the selection process works because there are checks and balances set up by the people at a variety of levels within the system. Thus while the power of the Pope seems to be absolute, it almost completely relies upon the information that has been collected by the pro-nuncio and the American Catholic Church.[10]

Can the selection process for finding a suitable bishop be improved? In 1972 "new" Vatican guidelines were published. The head of the Canon Law Society of America contended "that the consultation process, although more liberal, was still too restrictive, and did not seems to reflect the awakening consciousness of the responsible people of God in the Church."[11]

Some people desired to strengthen the part that the American bishops have in process of selection. Other critics wanted to give a greater voice to the clergy and also to the laity. Finally, some canon lawyers want the part played by the Bishops to be increased. For example, the Canon Law Society of America contends that every diocese should have a group or committee for the selection of priests for the position of bishop that would find nominations from different Catholics. One position is for an eleven-member committee which would be chosen by the bishop; the other members would be selected by the diocesan pastoral council.[12]

After reviewing the selection and the condition of the diocese, the committee would send its report and the list of choices to the priests' council. The report and the selections would be sent to the diocesan bishop who would also make his own investigation. During the spring bishop's conference the candidates would be rated on by these bishops. The voting results would be sent to the National Conference of Catholic Bishops' Committee on the selection of bishops which would close the group of candidates for the office of bishop for a specific office. The Pope would curtail or limit itself to three or five nominations sent by the NCCB Committee on the selection of bishops.[13]

Vatican leaders and many bishops are opposed to making the selection process more open and democratic. This would make it too "political." Opponents contend that the selection process is political anyway and that the selection process is run by cliques in Rome. A flaw in the selection process is that many priests do not take it into account seriously. The priests can have an input by recommending some of their fellow priests to the bishop and the papal pro-nuncio and by filling out the questionnaires on the candidates. Lay people can describe the type of bishop they would approve of to be their leader.[14]

The Bishop's Agenda

The priesthood is voluntary. What are a bishop's beliefs? While there are many theological disagreements in the Church, its participants agree on many of its beliefs, and they share a similar liturgy and sacramental system; and, they share the same Bible. The wisdom and books of the Second Vatican Council are also accepted by

A CATHOLIC AMERICA--A MATHEMATICAL PERSPECTIVE

many. The post-Vatican II documents emphasize "service, collegiality, the principle of subsidiarity, and the pastoral role of the bishops" has guided episcopal behavior. The idea of subsidiarity says that civilization in the Church or society should be developed at the lowest level.[15]

Andrew Greeley found in his research that Catholics in America favor by 61 to 39 per cent, that Church leadership should be delimited to the "bishops, priests, and deacons." Many Catholic priests support episcopal government. However, Father Greeley found that 29 per cent of the clergy find episcopal governance to be a problem for them. They desire a decentralized church power. They want, for example, a greater say for the priests'senate and a greater say for the laity.[16] Recently there has been a decline in priests' dissatisfaction to 22 per cent. Why did this happen? Many alienated priests have left. And, the Ecclesial environment has become less authoritarian and more pastoral. Also the bishops have developed a more consultative style of leadership.[17]

"Canon Law"

Canon law defines the legal system that an archbishop works within. It sets forth his power. Each seminarian takes course work in canon law. There is usually a priest on the bishop's staff who is degreed in canon law. A bishop, for example, "cannot rewrite liturgical texts or ceremonies, ordain women or married men, or consecrate a bishop without Vatican approval." Yet the bishop has a large amount of discretion in running his diocese. There is not a separation of power as we have it in America. The bishop is judge, executive, and legislator in his diocese or archdiocese.[18]

As a legislator the bishop has outlines for people under him in his diocese. The bishop must also have a group of consultators drawn from the priests council. He also has a finance council. The bishop is, furthermore, the chief executive. He can develop policies, programs, and develop offices. The bishop can fire a number of his administrative personnel. He appoints his own vicar general and he appoints a chancellor who often acts as the head of staff. The chancellor is the guardian of the archives and the chancellor helps prepare canonical statements.[19]

During 1983 the Code of Canon Law provided for a new diocesan office, the "moderator of the curia." This moderator must "coordinate the exercise of administrative responsibilities; and, to see to it that the other members of the curia fulfill the offices entrusted to them." A new officer stipulated under the latest Code of Canon Law is the financial officer.[20]

As executive leader, the bishop selects chancery leaders and pastors. Finally, the bishop has a role as a judge. The diocesan tribunal is under the bishop, however, this office deals mainly with the annulment of marriages. This legal problem is usually left to the annulment tribunal and to the canon lawyers.[21]

Reese, the author of the book, Archbishop, argues that American archbishops spend little time learning the fine points of canon law. Instead, they spend it developing their archdioceses.[22]

The Diocese of the Archbishop

Often, nearly 84 percent of the time, the archbishop is from a different archdiocese. Reese, claims that each archdiocese is unique. For example, archdioceses have different densities of population, different geographic areas, and different numbers of Catholics. Usually the boundaries of the diocese are similar to political entities. One exception is New York City where the diocese antedates the political unity of Brooklyn and New York. Thus when New York City placed in its contracts a homosexual rights clause, the bishop of Brooklyn signed the contracts and the archbishop of the diocese of New York refused to sign the contract.[23]

What about geographic size? The area of an American archdiocese usually averages 17,700 sq. miles which is twice as large as the state of Massachusetts. What about population? The average archdiocese runs close to 695,000 Catholics. The archdiocese of Los Angeles has a population of 2.7 million Catholics. This is the largest archdiocese. The smallest archdiocese is Anchorage which has only 22,982 Catholics.[24]

Sometimes splitting an archdiocese makes it easier to run. Cardinal Timothy Manning of the archdiocese of Los Angeles spun off San Bernadino and also Orange counties. The usual archdiocese is placed in one of three different categories. First are the most heavily populated dioceses such as Boston, and Chicago which have a density of 300 or more Catholics "per square mile." When an archdiocese is focused on one city and has a high population density, it usually has a centralized and large chancery. During the 1970's, John Seidler discovered that the more urban an archdiocese, the more liberal the bishops were regarding the evaluation of the Church. But, large urban dioceses were more likely to have dissident priests.[25]

This group contains the dioceses with 90 to 190 Catholics in each square mile. These dioceses usually have a large urban area, suburbs, and a country area. Also, these archdioceses, for the most part, have a strong chancery and regional vicars.[26]

The final category has archdioceses with a population of 20 or less per square mile, for example, Anchorage and Atlanta. In these archdioceses chanceries are often small. An archbishop can travel large distances to visit rural parishes. In Omaha, for example, the archdiocese has nearly two thirds of its Catholics living within the metropolitan area, but nearly two-thirds of the priests are in the outlying area.[27]

Population Composition

The composition of the population in a diocese can influence it leadership. In Miami, for example, the influx of Cuban and Haitian peoples has increased the Catholic population dramatically. In Atlanta new parishes are being founded. The

A CATHOLIC AMERICA--A MATHEMATICAL PERSPECTIVE

ethnic composition of a diocese usually elicits a response with programs and offices. The Church has had to meet the needs of waves of immigrants such as "Irish, Italian, German, and Polish peoples." Presently, dioceses that have large numbers of Hispanic peoples, like Miami, have offices for their Hispanic peoples.[28]

In Los Angeles and San Francisco reside large numbers of Hispanics; and, there are also immigrants from Asian and Pacific countries. These immigrants require special sensitivity. In Anchorage the archbishop should be especially sensitive to Indians and Eskimos. In central-city archdioceses the bishop is called to deal with black problems.[29]

Dioceses have also created offices and programs that are gauged to occupation groups, for example, farming in rural dioceses like Denver and Omaha. In Miami the retired and senior citizens are a concern. In Washington and New York the diplomatic corp require excellent attention. Also, university students, Catholic hospitals and prisons need chaplains to meet their spiritual needs. The Church is trying to deal with the aids crisis. The poor people need attention. In Minnesota and Massachusetts the liberal political climate affects the Church.[30]

An archbishop has many "institutional commitments" such as private schools and Catholic hospitals to provide support for. Most Catholic colleges and universities are staffed by different religious orders. These schools can further the education of Catholic priests and laity. There can be conflict between the bishop and the Catholic theologians. Some cities such as "Chicago, Los Angeles, New York, and Philadelphia" have so many Catholic schools that they are larger than many public school systems in America.[31]

Catholic hospitals in the larger cities like "Chicago, Cincinnati, Los Angeles, New York and Detroit each treat well over 1 million patients every year." "Medicare, and Medicaid cuts, medical ethics, malpractice insurance, and hospital finances" often have to be dealt with by the archbishop.[32]

Some archdioceses support many social service programs such as nursing homes and food kitchens. Other archdioceses have a wide range of personnel. Many of these people are quite talented and dedicated. The archbishop has to meet their needs and pay their salaries. Religious women teach in schools and work in hospitals and in different parish ministries. The decline in the number of sisters has hurt the Church. Lastly, there are the parish priests who already are set in their parishes when the archbishop arrives. Their needs have to be met.[33]

An auxiliary bishop can really help out an archbishop who may be from a different archdiocese. The auxiliary is familiar with the people, personnel, and institutions of the archdiocese. Some archbishops are very difficult to follow like Cardinal Francis Spellman of New York and Cardinal Richard Cushing of Boston.[34]

Style and Substance

What should the ideal archbishop be like? Reese claims that the ideal archbishop should be like a shepherd who is sensitive to his flock and an excellent administrator. He must be able to preach the gospel without threatening his people; and, he must be able to develop large social services and education programs with a small budget. His personality and stewardship must be orthodox and collegial; and, he must keep his flock happy. He must help the poor; and, he must have his priests think that they are the greatest people in the diocese. Also, the archbishop must not alienate the laity by being too clerical.[35]

The archbishop should be a national and international leader in the Church. He must understand the world of finances; and, he must be loyal to the Holy Father. He must be ecumenical but not lose the Catholic identity.[36]

Most archbishops come from the working-class. Half of their fathers don't have college degrees. The average age of an archbishop when he is appointed is fifty-three years. He has to retire at age seventy-five. Sixty-one percent of the archbishops were diocesan bishops when they received their promotion. On the average they were already bishops for six years. However, a number of archbishops had not worked in a chancery before being appointed a bishop. Some priests who became bishops were rectors in the seminary. Lastly, the archbishops are more educated than many priests. Nearly 30 percent have a S.T.D. theology degree or a J.C.D. canon law degree. Most are workaholics.[37]

Public Figures

In their own communities archbishops are important civic leaders. Cardinal Terence Cooke, for example, gave the invocation at the first AIDS research conference. Archbishop Flores urged his San Antonio Catholics to vote.[38]

Ecumenism Supporter

For an archbishop ecumenical attempts are strong with Jews in Los Angeles and New York and with Protestants throughout the whole country.[39]

Trans World Leaders

Some archbishops have national and international impact. For example, Cardinal Krol helped deal with the Catholic Relief service, aid for Polish farmers, and Vatican finances. Cardinal O'Connor is a leader of the Bishop's Congregation. Cardinal Law is helping develop the new catechism. Looking at the national scene, the bishops choose a new president and vice-president of the National Conference of Catholic Bishops. These leaders usually act as spokespersons for the NCCB on many Church issues. These leaders also present the American views of the Church to Rome.[40]

A CATHOLIC AMERICA--A MATHEMATICAL PERSPECTIVE

Teaching

All archbishops are teachers. But only Archbishop Whealon of Hartford is currently teaching in a formal classroom. Cardinal Law, for example, expressed the view that Boston College could become more Catholic. Cardinal Hickey dealt with the Curran episode whose view on sexuality was challenged by Church officials. Most archbishops teach through their sermons. Many archbishops publish their views in the Catholic newspaper. None of the bishops have published an article in Theological Studies, America's most excellent theological paper. Episcopal pronouncements that have national impact are published in Origins. No archbishop has been as able in utilizing the mass media as have several Protestant TV evangelists. Archbishops should be pastoral. That is, they should get along with their fellow priests; and, they should keep in step with their parishes.[41]

Church Administration

For much of American history the archbishops were largely administrators. In Europe, for contrast, the bishops were not chosen for financial competence. Finances in Europe were handled by officers of "state recognized churches." No American archbishop has an M.B.A. Some archbishops like Cardinal O'Connor are continually developing new programs. Other Archbishops, like Cardinal Donnellan of Atlanta, try to maintain peace in turbulent times. Some archbishops develop a supportive style. For example, archbishop Roach, helped develop a program called Renew, a program for education and parish regeneration.[42]

Archbishop Roach and the Common Good

What are the Church's teachings on social and economic problems? Archbishop Roach enumerates six principles. The first principle is human dignity. People should count or take priority over objects. The second principle is that we should all hear a call to community. Mankind has a social dimension. Each person should live for the good of all. The third principle is the recognition of the rights and responsibilities of the person. People have a right to life and to the things that enhance life such as food, shelter, clothes, medical care, education and work. The fourth principle is to be found in the workplace. Workers have rights and the work performed has dignity. Christ was a worker. The fifth principle is the sanctity of the poor. The poor need to be empowered. The sixth principle is solidarity. All mankind should work for peace; and, we all should work in solidarity to improve the planet and the environment. We should all work to solve the human equation of charity. People are more important than numbers or experiments.[43] Furthermore, the Archbishop is opposed to racism. He says that "in religious terms racism is a sin, a radical evil that divides the human family. It is a sin that violates human dignity in a very basic way."[44]

The United States Economy: A Catholic Bishops' Approach

Why do the Bishops write? They write as heirs of the prophets of the Bible who call us to walk with humility with God (Mi. 6:8); and, they write as disciples of Jesus who instructed us in the Sermon on the Mount that **"Blessed are the poor in spirit...Blessed are the lowly...Blessed are those who hunger and thirst for justice...You are the salt of the earth...You are the light of the world..."**(Mt. 1-6, 3-14). Within the parable of the last judgment Jesus stated that **"I was hungry and you gave me to drink...as often as you did it for one of these the least of my brothers, you did it for me."**(Mt 25: 35-40).[45]

Dominant Themes of the Bishops' Pastoral Letter

First, all economic judgments should be evaluated on the basis of whether it guards or undermines the dignity of the person. The dignity of the person is holy. Second, this dignity can be gained or guarded only by the person's community. The person is social. Third, everybody has a right to engage in the economic life of the society. Fourth, society has an obligation to the needy. Fifth, people should have human rights--civil, political, and economic rights. Sixth, society has the moral obligation to protect human dignity and to protect human rights.[46]

The Bishops, also, support efforts to stabilize the family in the economy. For example, they back efforts to stop the loss of family farms. The bishops think that we should stop the concentration of agricultural ownership. Moreover, they contend that poor nations should be defended; and, the bishops affirm the church's pronouncements on worker's rights, such as "collective bargaining, private property, subsidiarity, and equal opportunity."[47]

A New Drum Beat

In their letter the bishops contend that the day has come for a "New American Experiment" to put into action economic rights, to expand the sharing of economic power and evaluate the success of this experiment by how far it expands the common good. Catholics should listen to a call for activism and conversion.[48]

Peace and Economic Justice for Mankind

Every economic point of view that is human, moral, and Christian must be guided by several questions. First, how does the economy act for its people? Second, what does it enable its people to do? Third, how do the people engage in it?

A CATHOLIC AMERICA--A MATHEMATICAL PERSPECTIVE

Today the United States economy is the most powerful economic force in the world. It has provided an unprecedented standard of living for many of its players. But, this nation's experiment has faced serious conflict and pain. For example, this nation was conceived with a lack of justice to the American Indians. It gained independence with a bloody revolution. Slavery left a terrible scar on the economy of our nation; and, it was only stopped by a horrible Civil War. Women's suffrage, industrial worker protection, the end of child labor, the efforts to bring the Great Depression to an end, and the Civil Rights struggle of the l960's all worked to transform the economy and the political institutions of the country. Today the U.S ethos emphasizes economic freedom. But, this ethos also set forth the idea that the market place should be safeguarded by human rights. Slaves, for example, are not to be traded.[49]

The natural resources of this planet were given to us by God for the use of all. Mankind holds them in trust. Today there is much poverty in our country. The poor account for 33 million Americans. There are another 20-30 million needy Americans. Children are the greatest single poor group. What is needed today is moral vision.[50]

The Church has a strong Biblical outlook. The strong points of Israel's belief are "creation, covenant, and community." These ideas give us a basis for reflection on such issues as economic and social justice. What are these Biblical perspectives? First, God created man in his own image. Because of this every person has an inalienable dignity that marks human life before the separation of the person into races and nations and exists before human labor and human achievement. Second, we are a people of God's covenant. The various laws of this promise work to safeguard human life and property and give us respect for the disabled in the community[51] Third, Jesus brings God's reign and justice. He tells his followers not to focus on materialism. Instead, we should look to God and his justice. Fourth, Jesus summons us to be his disciples in his community. We should initiate God's life in community. Fifth, we have to deal with the problem of poverty and wealth and heed Christ's first public announcement: "**the spirit of the Lord is upon me to preach the good news to the poor(Lk 4:18 cf. Is 61:1-2).**" Sixth, we should build a community based on hope. Seventh, the Church has a tradition that lives. A message on U.S. economic life should be based on the idea of the Kingdom and on discipleship; but, it should be shaped also by Catholic life and reason. The indication is that our economy should be based on ethical values.[52]

Social Living has a Number of Responsibilities

What are these responsibilities? First, Christian people should work towards solidarity with the suffering; and, second, they should work against injustice. Third, Christians should work to include marginal people in our society. People have basic human rights such as "freedom of speech, worship, and assembly." People also have economic rights such as the "right to life, food, clothing, shelter, rest, medical care and education." People have a right to fair employment, education, and yes, social security. The Church is for the rights of laborers to develop labor unions. Also, Pope John Paul II points out that business and finance should receive protection but they should be held accountable to the common good. The Catholic Church has long supported the right to private property. However, it is not the Church's position to argue for a specific economic system. But, the Church thinks that the right to employment is a basic right. The Church opposes discrimination against women.[53]

Catholic social preaching does not argue for complete equality of income. Some inequality is necessary to deal with the problem of incentives and risks. But, "Whatever belongs to God belongs to all," says Pope John Paul II.[54]

"Food and Agriculture"

The acid test of a farm economy is its capability to meet the needs of present and future generations with justice. First, today farmland ownership is becoming more and more concentrated. Second, farm towns are decaying as people move to the city. Third, our resource base is being depleted through topsoil erosion, chemicals, and underground well's depletion. Fourth, racial minorities are often excluded from the farms. Farmers need solidarity.[55]

The Global Economy

Half of the people of the world live in abject poverty. The Church argues for "the preferential option for the poor." The Church urges the United States to work towards a global common good. However, the United States cannot be the sole provider or savior of the Third World.[56]

Wanted: A New Equation for the American Economy and the World Common Good

Over 200 years have gone by since the United States began its experiment in democracy. The founding fathers set forth to create justice, defend the American general welfare, and build a civilization based on liberty. Much work is left unfinished. To complete this task, new means of partnership and cooperation in the economy must be developed. The economy should be made more democratic. Cooperative work efforts are needed to make local, national, and international economic efforts more just. The Church has developed what it calls the "principle of subsidiarity" which highlights small or medium size communities or institutions developing moral responsibilities. This "principle of subsidiarity" asks for governmental intervention when smaller units of the economy are unable to promote justice. Partnership is needed to implement national policies.[57]

A CATHOLIC AMERICA--A MATHEMATICAL PERSPECTIVE

The Future

The development of social bodies begins with the conversion of the heart and soul. Work and prayer need to be united. Each one of us should heed a call to sanctity in this world. A strong family unit needs to be fastened by the economy. There must be a renewal in the Church's economic life along the lines of "(1) wages and salaries; (2) rights of employees; (3) investments and property; (4) works of charity; and (5) working for economic justice." What is the Church's prescription for the U.S. economy? It is a covenant of love and justice and freedom.[58]

A Bishop's Pastoral: Archbishop Roach on Sexuality

Parents have an important role in teaching their children about sexuality. What do we mean by sexuality. It is God's gift to mankind. Catholics think that the sacrament of matrimony unites a woman and a man in their quest in life. Matrimony transcends a couple's relationship. It follows from this that every conception has a right to a loving family, to a role in the community, and to God's graces. Teenagers need special help to integrate their "emotional, physical, spiritual, and psychological development." This is a process that continues on throughout all human life. Archbishop Roach contends that we need a new consensus that teens should abstain from sex. Knowledge does not always lead to virtue. Contraception and abortion are not allowable methods of controlling birth. They devalue life. Casual teenage sex often leads to poverty. Teenagers are not fully mature. Parents should set limits on teenager growth.[59]

What can be done? Married people can be faithful. Fidelity gives a larger meaning to sexuality. The young should learn to put fidelity on a pedestal.[60]

What should parents do? First, they should think that they are good models. They should try to discover what values are important to them. Second, the parents should respect a young person's experience of first love.[61]

What are some tips for parents? First, you should try to be approachable. Second, you should be affectionate and loving to your partner and offspring. Third, you should set limits on sexual behavior. Fourth, you should build appropriate ways to express tenderness. Fifth, you should become familiar with parish resources.[62]

What should teenagers do to be chaste. First, teenagers should understand that pre-marital sex is wrong. Sex belongs within the sacrament of marriage. Second, parents should be open to teenagers about their values especially abstinence. Third, abstinence means that teenagers should not develop exploitive relationships. Fourth, by postponing sex until the youth is married, he or she will not have to worry about pregnancy and abortion. Fifth, through abstinence and fidelity, venereal diseases such as aids can be avoided by teenagers. Sixth, by focusing on chastity, the teenager will have time to develop their relationships with important

57

people in their lives. Seventh, the teenager will not become fixated at an early stage of development and miss out on the richness of adult life. Finally, parents and teenagers should take an activist stance on Christianity and sexuality. Patrick's advice is that you should take life seriously.[63]

Archbishop John Roach on Spirituality

"Holiness is Everyone's Business."
Archbishop John Roach

What is holiness? It is a call to be a genuine friend of Jesus Christ. We must all heed "a call to holiness." Archbishop John Roach recalls that during his teens he felt attracted to the priesthood. He says that he felt that he did not have a personal holiness. Archbishop Roach recalls struggling with confession, however, he had a parish priest with whom he developed a strong attachment towards. Archbishop Roach learned that God loved his holiness or lack of it. He thought of "equating holiness with perfection." But he knew that he was imperfect. Archbishop Roach resolved this dilemma by surrendering his will to God's will. By doing this he realized that he was Christ's friend and that he was holy.[64]

Holiness is a call to community and worship. It is a call to service. It is a call to ministry. It is everyone's call. A call to holiness is also a call to work for justice. Christ is our friend. We can enhance our friendship with Christ by attending Mass and the Sacraments and by finding a "spiritual director."[65]

Grass-Roots Administrator or Clerical Monarch

The expansion of archdiocesan programs following Vatican II made a monarchical approach for an archbishop more taxing. In small dioceses, an archbishop can be enveloped in running different agencies. However, some archbishops tried to be royalists in large dioceses. In Chicago, for example, Cardinal Cody personally signed all the archdiocese checks for the city of Chicago.[66]

Delegation of Power

In larger archdioceses it is necessary to delegate responsibility or control will halt. Differences in the style of administration are seen most readily when a new archbishop comes into control. In New York, for example, Cardinal Cooke was almost always involved in making decisions. In contrast, Cardinal O'Connor delegates much of his authority.[67]

A CATHOLIC AMERICA--A MATHEMATICAL PERSPECTIVE

Model of Regional Vicariates

In heavily populated archdioceses the tendency has been to develop regional vicariates run by auxiliary bishops. Sometimes priests can be regional auxiliary bishops. "Baltimore, Boston, Chicago, Detroit, Dubuque, Los Angeles, Newark, St. Paul, and Washington have regional vicariates and are often separated into deaneries. Sometimes the archbishops group agencies into departments run by department heads. Joseph Bernardin, for example, when he was appointed archbishop of Chicago put his archdiocesan agencies under the control of six directors. The archdiocese of St. Paul and Boston modeled their administrators after the Chicago model.[68]

The first secretarial model was set up by Cardinal Francis Spellman for his New York archdiocese. Under this model seven administrators ran different parts of the Church's bureaucracy. They make reports to the archbishop. The typical departments are for finances, social services, and education. Some archdioceses have pastoral and ministries departments, for example, liturgy and marriage programs.[69]

Some agencies are hard to departmentalize, for example, youth ministry or the diocesan bulletin. Hispanic, black, and family agencies need a partnership with other diocesan programs. At times there is an agency for "leftover" agencies like the cemeteries' agency or the public relations agencies. In Boston, for example, Cardinal Law has fifty-four agencies under him. He reorganized these agencies into seven: "social services, pastoral services, education, community relations, health care services, ministerial personnel, and central services." Some priests have complained that the departmental system develops into a clique of bureaucrats who get in the way of the priests and the archbishop.[70]

"Vicar General/Moderator of Curia Model"

To coordinate diocesan problems an archbishop or vicar general can rely on themselves or the archbishop can select a "moderator of the curia." Often times this moderator is also a vicar general. He coordinates administrative problems and personnel problems.[71]

Chief Executive Model

In St. Paul archbishop Roach is the chief officer. He is a great administrator. He runs his own cabinets' get togethers. He has cabinet meetings every week and monthly gathering with the officers of the archdiocesan departments. Many priests, however, detest weekly meetings. Many archbishops have individual or small group meetings. The deceased archbishop Casey of Denver did not work easily in groups.[72]

A CATHOLIC AMERICA--A MATHEMATICAL PERSPECTIVE

Consulting Bodies

Consultative groups outside of the chancery have developed to be a critical part of Church leadership. Cardinal Bernardin works well with consultative groups. Canon lawyers emphasize that these groups are not deliberative. That is, "they do not make decisions unless the archbishop specifically delegates this power to them." In a way they are like "interest groups."[73]

"The College Consultors"

This type of college predates the Second Vatican Council. It is a "diocesan consultors group." This college is composed of six to twelve priests. They are appointed by the Bishop for five year appointments. The latest code of canon law stipulates that the consultors be priests. These consultors have a major role presently in the selection of an archdiocesan administrator who runs the diocese during the period between the death of an archbishop and the arrival of his successor.[74]

"Presbyteral Council"

The Second Vatican Council said to the bishops that they should have rapport with their priests. "In order to put these ideals into effect a group or senate of priests representing the presbytery should be established." "These priests would be collaborators of the bishop in the government of the diocese." "By the end of 1966 some 45 senates were functioning, and a year later 135." In 1983 the code of canon law developed laws regulating their membership and constitutions and switched their name to presbyterals'councils. Only about fifty-percent of the council were designated to be elected. However, the archbishop or bishop must consult the council on different laws: "the advisability of a synod; the modification of parishes; offerings of the faithful on the occasion of parish services; norms for parish councils; the construction of a church or the conversion of a church to secular use; and the imposition of a diocesan tax." The bishop is the head of the council and sets his agenda. Some councils, after dealing with priests finances, for example, salaries and medical bills, have had as their agenda abortion and South Africa.[75]

"Pastoral Councils"

Many archbishops have developed archdiocesan "pastoral councils" to bring about greater lay participation in the church. In 1933, nearly 36 percent of the dioceses had pastoral councils. "The function of this council will be to investigate and weigh matters which bear on pastoral activity, and formulate conclusions regarding them."[76]

A CATHOLIC AMERICA--A MATHEMATICAL PERSPECTIVE

"Archdiocesan Synods"

A particular consulting body that arose in the fourth century is the diocesan synod. It is "a group of selected priests and other Christian faithful of a particular church which offers assistance to the diocesan bishop for the good of the entire diocesan community." A Directory on the Pastoral Ministry of Bishops lists five methods a synod can assist the bishop: "(1) adopting laws and norms of the church universal to local conditions; (2) setting policies and programs of apostolic works; (3) resolving problems of the apostolate and administration; (4) giving impetus to projects and undertakings in the diocese; (5) correcting errors in doctrine and morals, primarily by providing authentic teaching."[77]

In spite of their shortcomings consulting bodies are excellent sources of advice and knowledge for different bishops. Different prerequisites are necessary for the consulting organs to be successful. First, there must develop a high level of partnership between the consulting persons and the archbishop. Second, the bishop must set forth critical issues to the members of the council. Third, the council must find someone to do secretarial work. Fourth, if a topic like education is on the agenda, the head of that department must be available. Fifth, council participants need training and orientation programs to evaluate their role in the diocese. "Robert's Rules of Order is not helpful for bodies desiring consensus rather than confrontation." Lastly, these councils need to communicate with their regional and parish level constituencies.[70]

Regional Rule

Different archdioceses have such a large number of parishes that it is not possible for the archbishop to deal with them directly. To deal with this situation, archbishops have divided some of their archdioceses into "regional vicariates" run by episcopal vicars or auxiliary bishops. The diocese can function also split off into deaneries. Newark, for example, is the smallest archdiocese in area but it is one of the largest in population magnitude "(1.3 million Catholics)." It is split into four vicariates with each vicariates led by an auxiliary bishop.[79]

"Power and Influence"

The powers a vicar has differ among archdioceses. Different vicars have access to the archbishops decision making. They may make themselves heard by becoming a member of the priests' personnel department or the finance board. Most archbishops think that visiting different parishes is a crucial part of their ministry. He can act as a model for priests and parishioners. Archbishops have had to come to terms with population changes that have emptied many inner city churches. In Detroit, for example, forty-three churches were closed in the inner city in one year.

A CATHOLIC AMERICA--A MATHEMATICAL PERSPECTIVE

To deal with the declining number of priests, some archdioceses are letting lay people administer the parishes. In Baltimore, for example, in 1988 the archbishop allowed nine of its 153 parishes to be led by lay administrators. Also, an archbishop has a number of agencies sponsoring programs for the diocese's parishes. A program that has met with approval with many archbishops is Renew. It is a faith renewal program first developed in Newark.[80]

Church Finances

Each archdiocese is a million or more financial unit. Anchorage, for example, is the smallest archdiocese's financial unit; yet, it has a 1.5 million dollar operation for its main department. In contrast, the income of the New York archdiocese surpassed $264 million in 1983.[81]

To assist the archbishop in his finances is the finance administrator. His office, the finance office, collects incomes, manages investments, pays bills, and negotiates salaries. To evaluate an archdiocese's finances it is necessary to differentiate between parish finances and the finances of the archdiocese's administrations departments. Parishes are almost completely supported by parishioners. The typical Catholic gives $320 annually. Dioceses are defined in civil law in different manners. Many dioceses in America, for example, Atlanta and Chicago are handled as a "corporation sole" whereby every parish is a separate "incorporated nonprofit corporation with the bishop and his administrators as trustees or members of the board of directors." High schools, seminaries and Catholic Churches are also "separately incorporated."[82]

Budgets

Every archdiocese stipulates that the pastor presents a financial statement to the archbishop annually. Mostly, these statements indicate the income and outflow of each parish for the economic year. A number of archdioceses request that the pastor draw up a budget that projects income and outgo of expenses for the coming fiscal year. Some parishes require money from the archdiocese; and, the finance administrator finances parishes with monetary problems. Several archdioceses, for example, Baltimore and Detroit have auditors that determine the reality of a parishes financial accounting. If fiscal problems are found, the auditor educates the pastor and his workers as to the nature of their parishes' problems. In certain archdioceses, like Washington, the archbishop must sign a contract for sums larger than $10,000. Large parish projects also need the archbishop's approval. For very large projects, the archdiocese often requests bids. Also, many archdioceses have a banking system that encourages parishes to deposit savings in their chancery. Sometimes new parishes have to be built. These new parishes need economic sustenance. Larger archdioceses provide monetary services such as medical insurance or worker's compensation, or insurance. Once in a while there is one big archdiocese's collection for many programs. In St. Paul this collection is called the "annual archbishop's appeal." It is collected on specified Sundays in he archdiocese's parishes. Many of the archdiocese's programs are budgeted. After a budget is submitted it is often subjected to percentage "increases and cuts." Sometimes a large debt has to be paid off. For example, archbishop Gerety had a $25.5 million dollar debt in his Newark archdiocese.[83]

Canon law stipulates certain "financial restrictions on a bishop." For example, a bishop must get the thumbs up of his finance department "before alienating church property worth over $500,000." Alienation is a technical term that means the transfer of the ownership of property from one person to another."[84]

"Personnel"

According to the archbishop, the critical personnel problems relate to the archdiocesan priests. It is the most crucial task in the archdiocese. There is a "shrinking" personnel supply. One study finds that there will be "only half the number of active priests by the year 2000 as there were in the I960's." Also, bishops are often perplexed about priests who leave their ministries, or have some difficult personnel problems. The priests' appointments to parishes and to archdiocesan departments is the major job of the bishop. To deal with the problems of priest's placements, many archbishops have set up a personnel board that is oftentimes elected. A smaller number of archdioceses lack personnel boards. In St. Paul, for example, a letter is sent out during January asking each priest if they want to move. The present tendency has been towards open listing whereby every diocesan priest is informed of an opening.[85]

"Regional Vicars"

Another troublesome area of priests' assignments is the area of the regional vicar. How does the personnel board meet the needs of the regional vicars and auxiliary bishops. In certain archdioceses, like Detroit and Washington, the auxiliary bishops sit on the staff of the personnel office. They take part in the board's decisions. A problem arises with the question of just how much confidentially should be maintained in personnel decisions. Privacy weakens when a particular priest is asked about a specific job.[86]

Another controversy is whether the archbishop should attend the meetings of the personnel departments. Some priests contend that the archbishop should not make decisions so that the archbishop can decide on appeals from priests who disagree with the personnel board. However, many personnel board members think that it is more efficient to have the bishop on the board.[87]

In priest's assignments it is important to distinguish between "associates and pastors." A recently ordained priest is seldom made a pastor. He is typically assigned as an assistant pastor so that he can work with a pastor. Depending on the diocese, a priest can remain an associate "for one and a half (Santa Fe) to twenty-five years (Boston)." Seniority is also a factor in the assignments of a pastor. If seniority is not taken into consideration feelings can be hurt. Many archdioceses' pastors have a six year term assignment.[88]

Locating the right priest for the particular parish is the aim of the personnel office and the bishop. Parishes can be large and urban or small and rural. In some parishes knowledge of Spanish is a consideration. Some parishes are liberal; others are conservative. Today a pastorate is a great responsibility. The objectives of the priest are an important consideration. Smaller parishes are often preferred. Schools can be a headache.[89]

After analyzing the parishes and the supply of priests, the personnel office makes its recommendations to the archbishop. The personnel board agrees that the archbishop has the final decision. Often he takes nearly 95 to 99 percent of the personnel office's recommendations.[90]

By canon law a priest should obey his bishop and he usually must take the parish that the bishop picks for him. However, if a priest says no, the bishop will, for the most part, acquiesce. When more than one priest desire a parish, only one can get the position. The problem of rejection is the most difficult part of the open listing procedure. Some archdioceses evaluate a priest's performance.[91]

"Comprehensive Personnel Office"

For secular and religious personnel, the bishop or the archbishop acts to determine personnel policies like salaries and benefits; but, they are not involved in assignments except for particularly visible administrators. For example, the leaders of the National Association of Church Personnel think that a comprehensive ideal is the way to approach "church personnel systems."[92]

In conclusion, the archbishop's personnel department typically begins with the finance office which deals with wages and benefits such as medical or retirement. Depending on the archdiocese it can develop into a sizeable office that screens job applicants, determines salaries, or provides for grievance procedures etc.[93]

Catholic Schools

"There are 7,659 Catholic elementary schools, 1,391 high schools, and 233 colleges and universities in the United States." Many of the colleges and universities are controlled or run by religious orders such as the Benedictines. But, 60 percent of the secondary schools and almost all of the primary schools are under local bishops. Today the majority of Catholic grade school students are in secular public schools. There are, however, 3 million primary school students in parish run religious programs.

A CATHOLIC AMERICA--A MATHEMATICAL PERSPECTIVE

The bulk of these students are educated by volunteers. The archbishop is responsible almost entirely for the dioceses' education programs.[94]

Running the Schools: "Archdiocesan Offices"

Many archdioceses have a separate office for the Catholic schools and for religious instruction programs. Some archdioceses such as Chicago and New York have a separate office for education. It is called the vicar of education or an education superintendent. The archbishop may also have under his supervision the seminaries and the campus ministry. In other archdioceses such as Atlanta or St. Paul, the secretary of education is the superintendent of the school system. Many archdioceses have an office of education that gives advice to the archbishop on educational policies. Particular archdioceses such as Baltimore or Cincinnati have a single board for schools and for religious teaching.[95]

The office of religious education runs the religious programs in each parish. There are approximately five thousand religious education directors in the United States. They often have a master's degree in sacred studies.[96]

Governing Schools

Schools are a very complex ministry. Policies have to be determined on a wide selection of issues relating to students and faculty such as admissions, tuition, firing, insurance etc. "The Church's policies and practices of governance and accountability are neither uniformly defined or universally practiced in Catholic schools," says Lourdes Sheehan director for the National Association of the Boards of Education for the National Catholic Education Association. Some archdioceses are less centralized than other archdioceses. Laity involvement is a decentralizing force.[97]

"Catholicity"

The focus of the Catholic school office is the supervisor who builds the school's academic and religious education programs. In addition to educational instruction, "liturgy, prayer, and Christian service" form a major part of a Catholic's life. Problems that are to be dealt with are ministry, student life, and teen pregnancy.[98]

Money

Finances are always a significant problem for the Catholic schools. Teacher's pay is sometimes set by the particular school and sometimes by the particular diocese of the archdiocese. Tuition covers 43 percent of the Catholic schools' revenues. Oftentimes tuition is less than day care. Many parishes subsidize their schools. When the schools have money problems, they are usually closed or merged.[99]

A CATHOLIC AMERICA--A MATHEMATICAL PERSPECTIVE

The principal and the pastors are the two critical participants in the parochial school system. The archbishop must be informed on problems in the schools. He can have a great impact on the schools. For example, while many Catholic schools are closing, a few archbishops such as "(Hanover of New Orleans and Strickes of Kansas City)" are building new high schools.[100]

"Catholic Social Services"

The Church runs a large number of social service programs for the "poor, sick, hungry, homeless, handicapped, emotionally disturbed, unemployed, teenage runaways, unwed mothers, battered women, abused children, refugees, alcoholics, drug addicts, prisoners, victims of AIDS, and others in need." The impact of these programs is enormous. Some of these programs occur in "hospitals, nursing homes, orphanages, and low-cost housing units." Other programs are handled out of shelters, foster homes or day care centers.[101]

When these programs are taken together the Church is the largest "nongovernmental provider of social services in the United States." For example, during 1987, 646 Catholic Charities gave services to 8.7 million people. Some Catholic programs are controlled by professionals. Volunteers staff others. Some programs have both professionals and volunteers. Part-time volunteers, for example, work in soup kitchens or in the St. Vincent de Paul Society.[102]

Catholic Government

Catholic social services are developed in a number of ways. Hospitals are usually incorporated "with their own boards." These boards run the hospitals. The archbishop may or may not be included on the board.[103]

Das Kapital

Throughout the nation, Catholic Charities in 1986 received $600 million. The greatest amount, 45 percent, comes from the government in the form of grants. Church sources account for 20 percent and service fees account for 17 percent. In St. Paul, for example, Catholic Charities received only "$800,000 of its $10 million dollar budget from the archdiocese." In Philadelphia, the Catholic Charities drive gets nearly $6 million from the separate parishes and $2 million from the United Way. Some hospitals, nursing homes etc. receive fees from the recipient: an insurance company, "Medicare, or Medicaid." Some programs get government grants. Several states purchase Catholic Charities' services. Low interest laws and grants finance many nursing homes.[104]

Are Catholic social services losing their Catholicity? Certain social service programs dilute their religious orientation to get money from outside the Church. For example, one archbishop was asked to sign a statement that his religious values had nothing to do with the archdioceses' social services. He declined to sign the statement. The archbishop is the person who is completely responsible for a diocese's social services.[105]

The difference between an archbishop and a bishop is that the archbishop is a metropolitan which means the leader of an ecclesiastical province that contains several dioceses. This metropolitan has little control over the diocesan bishops in his province who are called or termed suffragans. A metropolitan can defend his suffragan bishops. For example, archbishop Borders of Baltimore opposed Rome's move for a visit or investigation of the Richmond diocese by a Vatican representative. Borders contended that if an investigation was necessary, it should be conducted by an American bishop. The investigation was finally completed by Archbishop May of the St. Louis diocese.[106]

Catholic State Conferences

Bishops meet together one or more times yearly as a "state conference which is often led by an archbishop." At this conference, the bishops draw up a consensus on religious policies and public issues. Pastoral letters have come forth from these conferences, for example, a California conference wrote a letter on AIDS. The Texas conference issued a pastoral on the spiritual care of Hispanic immigrants. Following the NCCB letter on the economy, "the Kentucky, Maryland, and West Virginia conferences issued pastoral letters on the economies in their states.[107]

NCCB/USCC

Every American bishop meets one or more times a year at the National Conference of Catholic Bishops and the U.S. Catholic Conference to decide on religious and secular issues of common importance. The prelates have a single vote per person except on financial matters on which auxiliaries can't vote. A president and vice president are elected to three year terms.[108]

The bishops have issued policy statements "on Central America, South Africa, the Middle East, Northern Ireland, the Church in communist countries, racism, capital punishment, health care, abortion, the Equal Rights Amendment, care for the terminally ill, food stamps, Medicaid, education, homelessness, international debt, immigration, tax reform, labor relations etc."[109]

A great amount of time and effort deals with Church issues, "like ecumenism, Catholic education, evangelization, family ministry, Hispanic ministry, catechetics, women in the Church, liturgy, and sacramental practice." Issues that are "binding" on the bishops require a two-thirds vote.[110]

Many bishops are involved with international problems. The bishops as a group consulted the bishops of South Africa before they wrote a proclamation on divestiture.

A CATHOLIC AMERICA--A MATHEMATICAL PERSPECTIVE

Moreover, the American bishops often work together with the Holy See to help govern the Church. Because councils are quite rare, a synod of bishops which offers advice to the Pope meets almost every three years.[111]

When a bishop is appointed he takes a loyalty oath to the Pope. At five year intervals, the American bishops visit Rome wherein they meet with the Pope. Ad Limina refers to the graves of Saints Peter and Paul that the bishops visit. The Pope gives and ad limina talk on issues like general absolution, abortion, and fidelity. During papal visits in 1979 and 1987, Pope John Paul II congratulated the American Church for loyalty "to the Apostolic See." He spoke of the "equal human dignity of women and their true feminine humanity." Between visits the Pope usually communicates by letter.[112]

In conclusion, the bishops' decision making is often reactive and it is not often proactive. Usually episcopal rule is incremental. What the American Catholic Church needs is a mathematical, or behavioral, or empirical approach to evaluate its programs. If the "underlying reality of the universe is mathematics," then behavior is based on mathematics.[113]

The Crux of Godel's Proof

Godel illustrates how to build a formula in arithmetic that represents the statement: "The formula G is not demonstrable." That is, it says for "itself that it is demonstrable." Godel also illustrates that G can be demonstrated only if its negative is demonstrable.[114]

Yet, if a formula and its negation can both be demonstrated then "G is a formally undecidable formula." That is, it is a correct formula of arithmetic. Because G is both true and undecidable, the axioms of arithmetic are not complete. We cannot logically derive all the truths of arithmetic from its axioms.[115]

Impact

The results of Godel's study are very far-reaching. Godel showed that the problem of finding for each and every deductive system, for example, a deductive system within which the entirety of arithmetic, an absolute consistency proof is quite unlikely. There are a myriad "number of true arithmetical statements which cannot be formally deduced from any given set of rules of inference."[116]

Godel seems to think that only a philosophical "realism" like Plato's can give us a sufficient definition of truth. What is Plato's realism? It is the idea that mathematics or philosophy do not invent their "objects" but often find them as Columbus discovered America. If this is accurate, then, the objects must "exist" prior to their particular discovery. Plato's idea is that the study of the objects of mathematics lives in an area that gives them access to the intellect alone.[117]Maybe logic is an extension of Catholicism. Life is incomplete and inconsistent. The Church addresses the human condition and values reason. It seems to the writer that Marxism, existentialism, and Christianity could be codified logically--a predicate calculus.

Thoreau's, <u>Walden</u>, is a statement about dissent, nature, race , and classicism. Skinner's, <u>Walden II</u>, is a statement about nature and behaviorism. <u>Walden III</u> is a statement about consent, mathematics, Catholicism and the universe.

Patrick's Final Thoughts

If Godel can prove that there are some things that are true but can't be proved to be true, then it seems that it can be argued that there are some things that are false but can't be proved to be false. Franey thinks that Patrick's thesis is false but he can not prove it. Maybe there are things that are false but can be proved to be true.[118] My mother, Patricia Kast, thinks that there are some things that are undecidable and incomplete in life, but, we are all a part of a larger whole.[119] This book ends on an undecidable note. Godel's proof is a paradox: it may be false but can not be proved to be false. The key to Catholic American research is mathematics and solidarity. Let God deal with the undecidable. Let God call America's hand. What do you get when you pull all of these Catholic themes together?-- **American Catholic Integralism!**

An Interesting Note

There are 774,746 words in the Old and New Testament. This is a composite number. There are 125,185 words in the Apocrypha. This is not a prime number.

Godel's Proof highlights the undecidable aspects of mathematics, psychology and religion.

Patrick's Name in Base Eighteen Numbers:

Patrick O'Dougherty = C0.[120]

A CATHOLIC AMERICA--A MATHEMATICAL PERSPECTIVE

BIBLIOGRAPHY

Berrigan, Daniel. The Discipline of the Mountain. New York: The Seabury Press, 1979.

_____. Lights on in the House of the Dead. New York: Doubleday & Co., l974.

Berrigan, Philip, S.J. Widen the Prison Gates: Writing From Jail, April 1970 - December 1972. New York: Simon & Schuster, 1973.

_____. A Punishment For Peace. New York: The Macmillan Company, 1969.

Cattell, Raymond, B. The Scientific Analysis of Personality. Chicago: Aldine Publishing Company, 1966.

Cuskelly. E. J., M.S.C. A Heart to Know Thee. New York: Paulist Press, l963.

Dylan, Bob. Complete Works.

Encyclopedia Britannica. Encyclopedia Britannica, Inc. University of Chicago, l979.

Fischer, James T. The Catholic Counterculture in America, 1933-1962. Chapel Hill, North Carolina: University of North Carolina Press, 1989.

Franey, Michael, Dr. A private conversation.

Hanley, Boniface, O.F.M. Hecker Mahwah, New Jersey: St. Anthony's Guild, 1983.

Hofstadter, Douglas, R. Godel, Escher, Bach: an Eternal Golden Braid. New York: Basic Books, Inc. Publishers, 1979.

Lewis, C.S. Mere Christianity. New York: The Macmillan Publishing Co. Inc., 1952.

Malloy, Joseph. Reverend. C.S.P. A Catechism for Inquirers. New York: The Paulist Press, 1976.

Nagel, E. and Newman, James R. Godel's Proof. New York: New York University Press, 1958.

National Conference of Catholic Bishops. "Economic Justice for All." A Pastoral Letter on Catholic Social Teaching and the U.S. Economy. Published by the St. Paul Seminary School of Divinity of the College of St. Thomas, St. Paul, Minnesota, 1987.

Niebuhr, Reinhold. "Christian Realism and Political Problems." Reprinted 1977 by Augustus M. Kelley. Fairfield, New Jersey: Charles Scribner & Sons, 1977.

O'Dougherty, James. My father.

O'Dougherty, John. My uncle.

O'Dougherty-Kast, Patricia, my mother.

Powell, John, S.J. Abortion: the Silent Holocaust. Allen Texas: Argus Communications, 1981.

Reese, Thomas, J. Archbishop. New York: Harper & Row, Publisher, Inc., 1989.

Reid, James, F. A Critique of the Political Philosophy of the Berrigans from the Perspectives of Augustine and Niebuhr. A Dissertation Presented to the Faculty of the Graduate School at the University of Missouri, Columbia. August, 1976.

Reher, Margaret Mary. Catholic Intellectual Life in America. New York: Macmillan Co., 1989.

A CATHOLIC AMERICA--A MATHEMATICAL PERSPECTIVE

Roach, John, Archbishop of St. Paul, Minnesota. "Reviving the Common Good." A Pastoral Letter. The Catholic Communication Campaign, 1991.
"A Friendship Unlimited," A Pastoral Letter on Spirituality." Archdiocese of Saint Paul and Minneapolis, 1988. "Grateful for the Gift: Sexuality, Parents and Teens." A Letter to Parents. Archdiocese of Saint Paul and Minneapolis.

Runyon, Richard P. and Haber, Audrey. Student Workbook to Accompany Fundamentals of Behavioral Statistics, sixth edition. New York: Random House, 1988.

Stone, Robert. B.F. Skinner. A Listen and Learn Tape. P.O. Box 2124 Reseda, California, 91335.

Thoreau, Henry David. The Annotated Walden and "Civil Disobedience," edited by Philip Van Doren. New York: Clarkson Potter, Inc., Publisher, 1970.

Walesa, Leck. Polish Solidarity Leader.

West, Beverly., Griesbach, Ellen., Taylor, J. and Taylor, L. eds. The Prentice-Hall Encyclopedia of Mathematics. Englewood Cliffs: Prentice-Hall, 1982.

A CATHOLIC AMERICA--A MATHEMATICAL PERSPECTIVE

I N D E X

A CATHOLIC AMERICA--A MATHEMATICAL PERSPECTIVE

A CATHOLIC AMERICA--A MATHEMATICAL PERSPECTIVE

A CATHOLIC AMERICA--A MATHEMATICAL PERSPECTIVE

A CATHOLIC AMERICA--A MATHEMATICAL PERSPECTIVE

A CATHOLIC AMERICA--A MATHEMATICAL PERSPECTIVE

A CATHOLIC AMERICA--A MATHEMATICAL PERSPECTIVE

Endnotes

1.   Dr. Michael Franey

2.   Dr. Michael Franey.

3.   Ernest Nagel and James R. Newman, <u>Godel's Proof</u> (New York: New York University Press, l958), p.6.

4.   Ibid., p. 10.

5.   Ibid. pp. 11-24.

6.   Ibid. pp.25-27.

7.   Ibid. p. 27.

8.   Ibid. pp. 27-32.

9.   Ibid. pp. 36-44.

10.   Ibid. pp. 58-63.

11.   Ibid. pp. 64-85.

12.   Ibid. p. 86.

13.   Ibid. p. 96.

14.   Beverly West, Ella Griesbach, Jerry Duncan, and Louise Taylor, <u>The Prentice-Hall Encyclopedia of Mathematics</u>, (Englewood Cliffs, New Jersey: Prentice-Hall, Inc., 1982), pp. 314-317.

15.   Ibid., <u>op. cit</u>. West <u>et. al</u>., p. 316

16.   Ibid.

17.   Ibid. pp. 316-317.

18.   Ibid. p. 317.

19.   Ibid. p. 119.

American Catholic Radicalism

1. A Critique of the Political Philosophy of the Berrigans from the perspectives of Augustine and Niebuhr. A Dissertation Presented to the Faculty of the Graduate School at the University of Missouri, Columbia by James F. Reid, August 1976, p.7.

2. Ibid. p. 8.

3. Ibid. p. 9.

4. Ibid. p. 10.

5. Ibid. p. 11.

6. Ibid. p. 13.

A CATHOLIC AMERICA--A MATHEMATICAL PERSPECTIVE

7. Ibid. p. 14.

8. Ibid.

9. Ibid. pp. 15-22.

10. Ibid. pp. 23-25. James O'Dougherty

11. Ibid. pp. 23-31.

12. Ibid. pp. 41-46.

13. Daniel Berrigan, The Discipline of the Mountain, "Dante's Purgatorio in a Nuclear World" (New York: The Seabury Press, 1979), et. passim. See also, Daniel Berrigan, Lights on in the House of the Dead (New York: Doubleday & Co., 1974), et. passim.

14. Daniel Berrigan, op. cit., The Discipline of the Mountain, p.14.

15. Ibid. op. cit. Reid, pp. 50-55.

16. Ibid. Section on Augustine, pp. 112-116.

17. Ibid. pp. 85-103.

18. Ibid. pp. 107-109.

19. Ibid. Reid, op. cit., pp. 112-116.

20. Ibid. pp. 118-122.

21. Ibid. pp. 126-132.

22. Ibid. pp. 145-147.

23. Ibid. pp. 145-147.

24. Ibid. pp. 149-151.

25. Ibid. pp. 155-158.

26. Ibid. pp. 155-158.

27. Ibid. pp. 168-170.

28. Ibid. pp. 173-174.

29. Ibid. p. 180.

30. Ibid. pp. 185-186.

31. Ibid. pp. 192-199.

32. Ibid. pp. 200-203.

33. Reinhold Niebuhr, "Christian Realism and Political Problems," Reprinted 1977 by Augustus M. Kelley (Fairfield, New Jersey: Charles Schribner & Sons, 1977), pp. 119-127.

34. Ibid. Niebuhr. op. cit., pp. 129-137.

35. Ibid. Reid, op. cit. pp. 203-212.

36. Ibid. pp. 217-232.

37. Ibid. pp. 233-234.

38. Ibid. pp. 237-256.

39. See James T. Fisher, The Catholic Counterculture in America, 1933-1962, (Chapel Hill, North Carolina: University of North Carolina Press, 1989).

40. See Ernest Nagel and James R. Newman, Godel's Proof, op. cit.. p. 10 and et. passim.

Thoreau: Walden I

1. Henry David Thoreau, The Annotated Walden and "Civil Disobedience" edited by Philip Van Doren, Stern Clarkson, and N. Potter(New York: N. Potter, Inc. Publisher) p. 456.

2. Ibid.

3. Ibid. pp. 457-458.

4. Ibid. pp. 458-459.

5. Ibid. p. 460.

6. Ibid. p. 461.

7. Ibid. pp. 462-463.

8. Bob Dylan.

9. Ibid. op. cit., Thoreau, pp. 464-465.

10. Ibid. p. 466.

11. Ibid. pp. 468-478.

1. B. F. Skinner, Walden II.

2. Robert Stone, B. F. Skinner, a Listen and Learn Tape, P.O. Box 2124 Reseda, Ca. 91335., Side one.

3. Ibid.

4. Ibid. Side Two.

5. Ibid. Side Two.

1. Reverend Joseph Malloy, C.S.P., A Catechism for Inquirers (New York: The Paulist Press, 1976), p. 35.

2. Ibid. p. 36.

3. Ibid. p. 37.

4. Ibid. p. 39.

5. Ibid. p. 45.

6. Ibid. p. 47.

7. Ibid. p. 50.

8. Ibid. p. 52.

A CATHOLIC AMERICA--A MATHEMATICAL PERSPECTIVE

9. Ibid. p. 55.

10. Ibid. p. 57.

11. Ibid. p. 60.

l2. Ibid. p. 62.

13. Dr. Michael Franey

1. E. J. Cuskelly, M.S.C., A Heart to Know Thee (New York: Paulist Press, 1963), pp. 1-8

2. Ibid. pp. 8-22.

3. Ibid. pp.23-24.

4. Ibid. pp. 33-187, et passim.

5. Ibid. pp. 187-207.

6. Ibid. pp. 208-222.

7. Ibid. pp. 237-244.

8. Ibid. pp. 255-282.

9. Drs. Watson and Crick discovered the double helix.

1. C. S. Lewis, Mere Christianity (New York: The Macmillan Publishing Co., Inc., 1952), pp. 17-22.

2. Ibid. p. 22.

3. Ibid. pp. 22-37.

4. Ibid. pp. 23-37.

5. John O'Dougherty quoting G. K. Chesterton.

6. Ibid. C.S. Lewis, op. cit. pp. 47-49.

7. Ibid. pp. 52-65.

8. Ibid.

9. Ibid. p. 115.

10. Ibid. pp. 171-190.

11. See Raymond B. Cattell, The Scientific Analysis of Personality, (Chicago: Aldine Publishing Company, 1966), et, passim.

1. Godel, pp. 68-70.

2. Ibid. p. 71.

3. pp. 75-76.

4. Ibid. pp. 77-78.

1. These numbers were derived from QUICKVERSE BIBLE CONCORDANCE. The address of this concordance is 375 Collins Rd. NE, Cedar Rapids, IA 52402. Dr. Michael Franey suggested some of these operations and numbers. The writer

A CATHOLIC AMERICA--A MATHEMATICAL PERSPECTIVE

suggested many of them also.

1. Margaret Mary Reher, <u>Catholic Intellectual Life in America</u> (Macmillan Co., 1989), p. 1.

2. Ibid. p. 2.

3. Ibid.

4. Ibid.

5. Ibid.

6. Ibid. p. 3.

7. Ibid. p. 6.

8. Ibid.

9. Ibid. p. 7.

10. Ibid. pp. 7-8.

11. Ibid. p. 9.

12. Ibid.

13. Ibid. p. 10.

14. Ibid.

15. Ibid. pp. 12-13.

16. Ibid. p. 16.

17. Ibid. p. 14.

18. Ibid.

19. Ibid. p. 15.

20. Ibid. pp. 15-16.

21. Ibid. p. 19.

22. Ibid. pp. 19-20.

23. Ibid. pp. 22-23.

24. Ibid. pp. 23-25.

25. Ibid. p. 25.

26. Ibid. p. 26.

27. Ibid. p. 27.

28. Ibid.

29. Leck Walesa came up with the idea of solidarity. Archbishop John Roach utilized the <u>Renew</u> program in his diocese.

30. Ibid. Margaret Mary Reher, <u>op. cit</u>., p. 29.

31. Ibid. p. 29.

32. Ibid.

33. Ibid. pp. 29-30.

34. Ibid. p. 30.

35. Ibid. pp. 30-31.

36. Ibid. pp. 31-32.

37. Ibid.

38. Ibid. p. 32.

39. Ibid. pp. 32-33.

40. Ibid.

41. Ibid. pp. 33-34.

42. Ibid. p. 34.

43. Ibid. pp. 34-35.

44. Ibid.

45. Ibid. pp. 35-36.

46. Ibid.

47. Isaac Hecker, quoted in Hecker: A Missionary to North America by Boniface Hanley, O.F.M. (St. Anthony's Guild, I983), Cover.

48. Ibid. p. 36.

49. Ibid. pp. 38-39.

50. Ibid. p. 39.

51. Ibid. p. 42.

52. Ibid. pp. 42-43.

53. Ibid. pp. 43-44.

54. Reher, op. cit., p. 55.

55. Ibid. pp. 45-46.

56. Ibid. pp. 46-47.

57. Ibid. pp. 47-49.

58. Ibid.

59. Ibid.

60. Ibid.

61. Ibid. p. 50.

62. Ibid.

63. Ibid. pp. 52-55.

64. Ibid.

65. Ibid. pp. 58-59

66. Ibid. pp. 61-62.

67. Ibid. pp. 63-66.

68. Ibid. pp. 66-71.

69. Ibid. pp. 72-75.

70. Ibid. pp. 76-79.

71. Ibid. pp. 85-88.

72. Ibid. pp. 88-90.

73. Ibid. pp. 93-94.

74. Ibid. pp. 94-97.

75. Ibid. pp. 98-105.

76. Dr. Michael Franey, in a private conversation. The examples of God's capability of killing himself or creating a stone he can't lift are high school philosophy ideas.

77. Ibid. pp. 105-107.

78. Ibid.

79. Ibid.

80. Ibid. pp. 112-113.

81. Ibid. pp. 114-141, et. passim.

82. Ibid. For a discussion of the Catholic American Counterculture see, James T. Fischer, The Catholic Counterculture in America, 1933-1962 (Chapel Hill, North Carolina: University of North Carolina Press, 1989).

83. See the New World Dictionary of the American Language for the definitions of fissionism, fusionism, and field theory.

84. See Richard P. Runyon and Audrey Haber, Student Workbook to Accompany Fundamentals of Behavioral Statistics, sixth edition, (New York: Random House), Table of Contents, and et. passim.

1. John Powell, S.J., Abortion: The Silent Holocaust (Allen Texas: Argus Communication, 1981), p. 3.

2. Ibid. pp. 5-10.

3. Ibid. pp. 16-32.

4. Ibid.

5. Ibid. pp. 32-33.

6. Ibid. 33-41.

7. Ibid. pp. 41-42.

8. Ibid. pp. 43-56.

A CATHOLIC AMERICA--A MATHEMATICAL PERSPECTIVE

9. Ibid. pp. 79-82.

10. Patrick O'Dougherty, and anonymous.

11. See the magazine, Just Life/90, JustLife Education Fund, 10 Lancaster Avenue, Philadelphia, PA 19151.

1. Claudia Chaves, a friend.

2. Thomas Merton.

1. Thomas J. Reese, Archbishop (New York: Harper & Row, Publisher, Inc., 1989), pp. Introduction.

2. Ibid. p. 1.

3. Ibid. pp. 2-6.

4. Ibid. pp. 7-9.

5. Ibid. p. 40.

6. Ibid. pp. 40-45.

7. Ibid. pp. 46-47.

8. Ibid. pp. 47-48.

9. Ibid. p. 48.

10. Ibid. pp. 48-49.

11. Ibid.

12. Ibid. pp. 49-50.

13. Ibid. pp. 50-51.

14. Ibid. pp. 51-52.

15. Ibid. pp. 53-55.

16. Ibid. p. 55.

17. Ibid. pp. 55-57.

18. Ibid. pp. 57-58.

19. Ibid. pp. 58-59.

20. Ibid. pp. 59-60.

21. Ibid.

22. Ibid. p. 61.

23. Ibid. pp. 62-64.

24. Ibid.

25. Ibid. pp. 65-66.

26. Ibid. p. 66.

27. Ibid.

28. Ibid. pp. 67-68.

29. Ibid. p. 68.

30. Ibid. pp. 69-70.

31. Ibid. pp. 70-71.

32. Ibid. pp. 71-72.

33. Ibid.

34. Ibid. pp. 72-76.

35. Ibid. p. 76.

36. Ibid.

37. Ibid. pp. 77-78.

38. Ibid. pp. 84-85.

39. Ibid. pp. 86-87.

40. Ibid.

41. Ibid. pp. 88-91.

42. Ibid. pp. 91-98.

43. Archbishop Roach, Reviving the Common Good (Published by the Catholic Communication Campaign, 1991), pp. 3-6.

44. Ibid. p. 17.

45. See the National Conference of Catholic Bishops, Economic Justice for All, Pastoral Letter on Catholic Social Teaching and the U.S. Economy, (St. Paul: The St. Paul Seminary School of Divinity of the College of St. Thomas), p. 4.

46. Ibid.

47. Ibid. pp. 4-5.

48. Ibid.

49. Ibid. p. 6.

50. Ibid. p. 7.

51. Ibid. p. 8-9.

52. Ibid. pp. 9-10.

53. Ibid. pp. 10-16.

54. Ibid. pp. 16-20.

55. Ibid. p. 20.

56. Ibid. pp. 20-24.

57. Ibid. pp. 25-26.

58. Ibid. pp. 26-28.

A CATHOLIC AMERICA--A MATHEMATICAL PERSPECTIVE

59. Archbishop John Roach, "Grateful for the Gift: Sexuality, Parents and Teens," A Pastoral Letter. St. Paul, pp. 1-5.

60. Ibid. pp. 5-6.

61. Ibid.

62. Ibid. p. 6.

63. Ibid. p. 6-7.

64. Archbishop John Roach, "A Friendship Unlimited," A Pastoral Letter on Spirituality. St. Paul, pp. 1-12.

65. Ibid.

66. Reese, op. cit., pp. 99-100.

67. Ibid. pp. 100-103.

68. Ibid. p. 103.

69. Ibid. pp. 104-105.

70. Ibid. pp. 105-106.

71. Ibid. pp. 107-108.

72. Ibid. pp. 109-111.

73. Ibid. p. 113.

74. Ibid. p. 114.

75. Ibid. pp. 115-118.

76. Ibid. pp. 118-120.

77. Ibid. pp. 123-124.

78. Ibid. pp. 123-125.

79. Ibid. pp. 125-133.

80. Ibid. pp. 135-147.

81. Ibid. pp. 150-151.

82. Ibid. pp. 151-152.

83. Ibid. p. 153.

84. Ibid. pp. 153-188.

85. Ibid. pp. 192-214.

86. Ibid. pp. 218-220.

87. Ibid. pp. 221-223.

88. Ibid. p. 224.

89. Ibid. pp. 232-233.

90. Ibid. pp. 240-245.

91. Ibid. pp. 246-248.

92. Ibid. p. 254.

93. Ibid. p. 256.

94. Ibid. p. 260.

95. Ibid. p. 261.

96. Ibid. p. 262.

97. Ibid. pp. 267-269.

98. Ibid. pp. 270-273.

99. Ibid. pp. 273-277.

100.Ibid. pp. 280-282.

101.Ibid. p. 284.

102.Ibid. pp. 284-287.

103.Ibid. p. 288.

104.Ibid. p. 291. Das Kapital is a book by Karl Marx and Frederick Engels.

105.Ibid. pp. 298-306.

106.Ibid. pp. 307-307.

107.Ibid. pp. 309-310.

108.Ibid. pp. 311-312.

109.Ibid. p. 312.

110.Ibid.

111.Ibid. pp. 313-315.

112.Ibid. pp. 316-331.

113.Dr. Michael Franey and Patrick O'Dougherty.

114.Ernest Nagel and James R. Newman, op. cit., p. 85.

115.Ibid. p. 86.

116.Ibid. p. 86.

117.Ibid. pp. 99-101. See also, Douglass R. Hofstadter, Godel, Escher, Bach: An Eternal Golden Braid (New York: Basic Books, Inc. Publishers, l979).

118.Dr. Michael Franey.

119.Patricia O'Dougherty Kast.

120.It was my idea to set up my name in base eighteen numbers. Mike Franey suggested the letters.